Droll Science

Droll Science

Being a Treasury of Whimsical Characters,
Laboratory Levity, and Scholarly Follies

Selected by

Robert L. Weber

Humana Press □ Clifton, New Jersey

Library of Congress Cataloging in Publication Data

Weber, Robert L. 1913—
    Droll science.

    Includes index.
    1. Science—Anecdotes, facetiac, satire, etc.
I. Title.
Q167.W418   1987   500   87-22642
ISBN 0-89603-112-8

Preface

It may be a pleasant surprise to the reader to encounter in this book a light-hearted side of science.

Off-duty scientists have frequently produced poetry, operettas, spoof biographies, parody reports, and cartoons inspired by their scientific work. Though originally offered chiefly for the enjoyment of their colleagues, many such occasional pieces can be read for pleasure not only outside of their original fields, but also outside of science itself—after all, the real subject of humor is always human nature *in flagrante derelicto!* We can thus be grateful that the many authors of the pieces printed here have been willing to share the fun with us in this anthology.

The anecdotes recounted in these pages that give insight into personalities, academic life, methods of research, and changing viewpoints in science almost certainly have the most lasting value. But some pieces are here unabashedly 'just for fun!' as Maxwell told Thomson who, surprised at viewing a little man on the stage of Maxwell's microscope, demanded 'What is *he* there for?!'

Robert L. Weber

Contents

People and Quirks

Condon's Filing System

Doctor Churchill Eisenhart had loaned a report to Dr. Edward Condon (National Bureau of Standards director from 1946 to 1951). Quite some time later he asked Condon if he could have it back. Condon replied, "I took it over to the manse. Go over and ask Emily, and she will find it for you." Eisenhart took Condon's advice, went over to the manse, and asked Mrs. Condon about it. She promptly escorted Eisenhart into the large living room, the walls of which were lined with bookcases and shelves. Approaching a partly filled shelf she turned and asked Eisenhart, "How long ago did you loan it to Edward?" He replied, "About two and a half months ago." Muttering to herself, "At a yard-and-a-half a month...." she measured off about three and three-quarters yards backward from the latest document and pulled over a "pinch" of documents about six inches thick, saying, "It is probably in here." It was.

NBS Standard
p. 4, 1980
US National Bureau of Standards

Eddington's Doorway

Scientific thinking taken to its extreme made it difficult for one distinguished mathematician and astrophysicist, Sir Arthur Eddington, to walk through his doorway.

I am standing on the threshold about to enter a room. It is a complicated business. In the first place I must shove against an atmosphere pressing with a force of fourteen pounds on every square inch of my body. I must be sure to land on a plank traveling at twenty miles a second around the sun—a fraction of a second too early or too late, and the plank would be miles away. I must do this while hanging from a round planet, head outward into space and with a wind of aether blowing no one knows how many miles a second through every interstice of my body. The plank has no solidity of substance. To step on it is like stepping on a swarm of flies. Shall I not slip through? No, if I make the venture one of the flies hits me and gives a boost up again; I fall again and am knocked upward by another fly; and so on. I may hope that the net result will be that I remain about steady; but if unfortunately I should slip through the floor or be boosted too violently up to the ceiling, the occurrence would be not a violation of the laws of Nature, but a rare coincidence. These are some of the minor difficulties. I ought really to look at the problem four-dimensionally as concerning the intersection of my world-line with that of the plank. Then again it is necessary to determine in which direction the entropy of the world is increasing in order to make sure that my passage over the threshold is an entrance, not an exit.

Verily, it is easier for a camel to pass through the eye of a needle than for a scientist to pass through a door.

Sir Arthur Eddington
The Nature of the Physical World
Cambridge University Press, 1928

The Tale of Schrödinger's Cat

Schrödinger called his cat and said,
"You can be both alive and dead,
For a linear combination of states
Postulates two simultaneous fates."
Poor shocked pussy could not say,
"I shall inform the SPCA.
Your pet theory seems to me
An ultraviolent catastrophy."

What then did this kitty do?
She looked at him and said "μ."

Marilyn T. Kocher
Physics Today
May, 1978

In honor of
Albert Michelson's
use of his interferometer
to measure the diameter of Betelgeuse

Professor Albert Michelson, Chicago, USA
Has found out things about the stars in quite a novel way,
Observing interferences of little waves of light,
Produced when homocentric rays are caused to reunite.
Of course, you do not understand what all these big words mean,
It's much as if I called the sky the empty empyrean;
It makes our simple little games appear more scientific,
To dress them up in words like these and not be too specific.
At all events, Professor M. with four small looking-glasses,
Can measure quite exactly the most distant stellar masses.
The little star that twinkled in our ancient nursery rhymes,
Is bigger than this sun of ours by many million times,
While the brightest one of all of those that sparkle in Orion
Is somewhat greater than our William Jennings Bryan.

And therefore let us make our bow in quite the proper way
To Professor Albert Michelson, Chicago, USA.

R. W. Wood
Contemptuary Science

==

Dirac Stories

Paul Adrien Maurice Dirac shared the 1933 Nobel Prize for physics with Erwin Schrödinger "for the discovery of new and productive forms of atomic theory"—quantum theory. On the occasion of his seventieth birthday, a symposium was held under NATO sponsorship at the International Center for Theoretical Physics, Trieste, Italy, 18–25 September 1972. The papers presented by luminaries of science occupy an 839-page book.

At the banquet of that symposium, some anecdotes were related of which the following are a sample:

Transmigration

S. Chandrasekhar: It finally became clear that the source of Fermi's illness was cancer.... When he asked, "How many more months?," Dr. Dragstedt replied, "Some six months."

The following day Herbert Anderson and I went to see Fermi in the hospital. It was of course very difficult to know what to say or how to open a conversation when all of us knew what the surgery had shown. Fermi resolved the gloom by turning to me and saying, "For a man past fifty, nothing essentially new can happen; and the loss is not as great as one might think. Now you tell me, will I be an elephant next time?"

Joseph Jauch: I had written a little piece in which I wanted to try to popularize some of the problems in the interpretation of quantum mechanics. It happened to be completed just at the moment when Professor Dirac and I were in Tallahassee together, and I thought this was a wonderful opportunity to present my essay to Professor Dirac and benefit by his comments. So

I left this piece with him, and a week later he brought it back to me and said, "Thank you." I was a little surprised that there was no more comment to it than just that. I prodded him a little bit as to what his opinion was about this piece, and he said, "I don't like the title of it." I asked, "Why not?" The title, incidentally was, "Are Quanta Real?" And then he said, "Well, it's just like asking, is God real?" And I said, "That is a very interesting remark, because this is just about what I wanted to communicate in this dialog." And then Dirac said, "Why did it take you so many pages to say it?"

Charles Enz: Dirac's new wave equation gave the correct value, 1/137, for the fine structure of the hydrogen lines, in which the de Broglie-Schrödinger theory had failed. This constant, introduced by Sommerfeld, had become a fetish for Eddington (in a cloak room, he would hang his hat on peg number 137) and it was Pauli's link to the "magic-symbol" world.

Pauli spent the last few days of his life in the Red Cross Hospital in Zurich, where he died on 15 December 1958. A fact which had disturbed him during these last days was that the number of his room was 137.

Paul Dirac: I would just like to say that this has been a very happy occasion for me. I have met so many old friends and every one of them has been a delight to meet again. I would like to thank you all for coming on this occasion. And I would also like to thank the various people who have spoken about me. I feel they have rather overwhelmed me with their nice remarks. They haven't talked at all about my failings, my forgetfulness, my absentmindedness.

Perhaps I might just mention a little story that illustrates to what extent absentmindedness can lead one. This concerns Hilbert, the mathematician whom all quantum physicists know about. He was perhaps the most absentminded man who ever lived. He was a great friend of the physicist James Franck and he said, "James, is your wife as mean as mine?" Well, Franck was rather taken aback by this question and didn't quite know what to say. He replied, "Well, what has your wife done?" And Hilbert said, "It was only this morning that I discovered quite by accident that my wife does not give me an egg for breakfast. Heaven knows how long that has been going on."

Before I forget to do so, let me thank you all very much.

Jagdish Mehra, Ed.
The Physicist's Conception of Nature
Reidel, Dordrecht, Holland, 1973

Starstruck

Fritz Houtermans was the first to work out the chain of nuclear reactions that provides the sun with its energy. On the day when the last piece of the puzzle fell into place, Houtermans had a date. His mind was full of the feeling that comes with the realization that one has solved an age-old mystery.

His Fraulein, as they walked the quiet, dark streets of Göttingen, sensing the exaltation of her escort's mood, remarked on the beauty of the night sky.

"Yes," the young theorist replied, "and would you have guessed that you were arm-in-arm with the only man alive who knows why they shine?"

R. H. March
Physics for Poets
McGraw-Hill, New York, 1970

Elsasser Noted

Perhaps the most reactionary and obscurantist notion that puts our society on the rock, so to speak, is the one held by so many supposedly educated people that the nitty-gritty of technology is beneath them, and that the clever manipulation of verbiage is enough.

Once when a new building was being erected at Caltech someone painted on the wood fence surrounding the site:

"Jesus saves." Two days later the students here had elaborated on this; it said underneath: "But Millikan gets the credit."

Walter Elsasser
Memoirs of a Physicist in the Atomic Age
Science History Publications, New York, 1978

Scientists in Double Dactyls

Rutherford

Pocketa-pocketa
Ernest Lord Rutherford
Blushed to confess, as he
Stared at the floor:

"Protons are vast, glory
Infinitesimal;
Sharing one's fame with a
Dane is a bore!"

Reichenow

Higgledy piggledy
G. Doflein-Reichenow,
Protozoologist,
      Said in a talk,

"Crass persons call the cal-
careous leavings of
Foraminifera
      By the name *chalk*."

Guggenheim

Higgledy piggledy
Edwin* A. Guggenheim,
Thermodynamicist,
Said, "Don't you see?

All you need know is im-
plied by dU equals
T times dS minus
P times dV."

Edwin? Well I don't know.

Clausius

Higgledy piggledy
Rudolf J. Clausius
Made two remarks that may
      Even be true.

One was an utterance
Thermodynamical:
"Entropie strebt einem
      Maximum zu."

Copernicus

Higgledy-piggledy
Nic'laus Copernicus
Looked at the Universe,
Spoke to the throng:

Give up your Ptolemy,
Rise up and follow me,
Heliocentrically
Ptolemy's wrong.

Arrhenius

Higgledy piggledy
Savante Arrhenius
Thought that the spores in a
    Vaporous sea,

Migrating planetward
Intergalactically,
Might produce beings like
    You and me.

Edison

Higgledy-piggledy
Thomas A. Edison
Turned on a switch with a
Wave of his wand,

Giving his name to some
Organization
Chaps in whose light we are
Now being Conned.

Berzelius

Higgledy piggledy
Jons J. Berzelius,
Tired of "Wasser" and
    "Water" and "l'eau,"

Said "I'm replacing this
Multilinguistical
Crap by the elegant
    Name, H_2O."

Alan Holden
Bell Laboratories, Murray Hill, New Jersey

Stokes' Clerihew

Sir George Stokes
Was not given to making jokes,
So the Fellows of the Royal Society laughed to excess,
When he said* "curl $\int F \cdot da = \oint F \cdot ds$."

*In 1854.

Andrade

    E. N. da C. Andrade, in his book *Rutherford and the Nature of the Atom,* tells a nice story about Julius Elster and

Hans Geitel, whose names are familiar to all physicists for their work on the photoelectric effect in the late 19th century. They did all their work in collaboration, and apparently were well known and recognized in their community. In their time, according to a story current in Germany, there was a man who much resembled Geitel in appearance. A stranger, meeting him one day, said: "Good morning, Herr Elster," to which he replied: "Firstly, I am not Elster, but Geitel; and secondly, I am not Geitel."

Contributed by **A. P. French**

Singer

Novelist Isaac Bashevis Singer, on his lack of aplomb with the scientific community: "I once wanted to make small talk with a physicist by asking, What's new in physics? But then I remembered—I don't know what's old in physics."

Anonymous

Hunting Dog

I hope that in a fairly long life, punctuated here and there by promotions of various types, that I have not reached the state of exalted position and complete uselessness that was achieved by one of the hunting dogs I heard about, trained by a northern woodsman. Their master, who had long enjoyed a warm acqaintance with a university community, had the habit of naming his dogs for faculty members that he admired. But when a few wives became a little indignant over the practice, he decided to name his dogs for various academic ranks—instructor, assistant, professor, and so on.

One hunting season, a man from Chicago hired for two dollars and a half a day a dog he liked very much. The following year, when he asked for the same dog, he was told the price would be five dollars a day. When he protested the steep inflation and insisted that it was the same dog, the owner agreed, but said that the dog had been promoted to assistant professor and was now worth the added money. The next hunting season, the price jumped to seven dollars and a half because the dog had then achieved the rank of associate professor, and the year after, it was raised to ten dollars, the reason being that the well-trained canine had reached the noble status of full professor.

The following year, when the hunter returned to rent the same dog, he was turned down. "Why not?" demanded the hunter insistently. "Well, I'll tell you," said the old woodsman, "I can't let you have him at any price. This spring we gave him another promotion and made him president of the college. Now all he does is sit around and howl and bark, and he ain't worth shooting."

Dwight Eisenhower
Rockefeller Institute
May 14, 1959

Half a Smoke Ring?

In a group of eminent physicists gathered in Geneva, R. W. Wood expressed some skepticism concerning the theory of vortices. The issue was narrowed down to the question whether or not it was possible to blow half a smoke ring, with Wood maintaining the affirmative against all the others. The climax came when he produced a cardboard box and thumped it with his finger. Half a smoke ring came out of an opening in the side of the box!

The explanation later disclosed that the box was partitioned inside, at the level of the opening, between an upper half filled with smoke and a lower half filled with a transparent gas heavier than air. Thumping the box produced a vortex having the doughnut shape required by the theory, but with only the upper half smoke-filled and visible.

Richard T. Cox
The Johns Hopkins University
September 16, 1981

Stories About Astronomers

Newton would often have forgotten to eat if he had not been reminded. On a certain day, Dr. Stukely was shown into the dining room where the noon meal was ready. Dr. Stukely waited a long while, became impatient, and took the cover off a dish on which lay a little rooster. He ate it, left the bones on the dish, and put the cover back on. A few minutes later Sir Isaac entered the room, and after saying hello seated himself at the table. When he took the cover off the dish and saw only bones he said, "How absent minded we philosophers can be! I thought certainly that I hadn't yet eaten."

* * *

Newton had a big cat and a little cat that he kept in a shed behind the house. He had a carpenter saw two doors in the shed: one big and one little one.

* * *

For Newton's breakfast the maid brought an egg that had to be cooked exactly five minutes, as checked by that gentleman himself. Once when she returned, there stood Newton attending the boiling water thoughtfully looking at the egg that he held in his hand while his watch lay in the boiling water.

* * *

Cavendish once sat beside Herschel at a dinner. Cavendish was a very taciturn man. Suddenly he asked, "Is it true, Dr. Herschel, that through your telescope you see the stars as disks?" "Round as a button," answered Herschel shortly. That was the end of their conversation.

* * *

Louis XIV visited the observatory at Paris to look at a new comet through the telescope. He asked Cassini in what direction the comet would move. This question could not be answered because no one had studied the comet long enough to know its course. Cassini smartly decided that the king most likely would not pay a second visit and so without hesitation or blushing he described an imaginary trajectory for the comet.

* * *

The observatory at Poelkowo was from the start one of the best establishments in the world, because the Russian government had expended on it an amount unheard of at that time.

When the Tsar came to see the completed institute he was conducted around by F.G.W. Struve, the first director. At the end of the tour, the Tsar asked, "Well, Struve, are you satisfied?" This very careful astronomer who did not want to deprive himself of further money answered, "For the time being, your Majesty."

* * *

Take note of all that shines.—*Motto of **William Herschel***

I would like very much to find a beautiful problem that is not difficult to answer.—*Maupertuis*

M. Minnaert
Translated by Colena Jordan from
De Sterrekunde en de Mensheid,
Katwijk, Servire BV, 1946

Independent Thinkers

Note that I say "independent thinker," and not "crank." The independent thinker is a genuine, well-meaning person, who is not hidebound by convention, and who is always ready to strike out on a line of his or her own—frequently, though not always, in the face of all the evidence. In some respects he or she is a rather special kind of person, though generally speaking, conventional enough except in one particular line of thought. He or she may or may not be scientifically qualified. All share the wish to inquire, and—this is the vital fact—all are anxious to do something really useful.

There are, of course, unconventional people who are decidedly different. Some religious (or pseudoreligious) cults come under this heading, and I think it is clear that they can do immense harm by influencing gullible people and even breaking up families. It is people of this sort whom I would label either as cranks or as crooks (sometimes both). The less I see of them, the more I am pleased.

So, let us concentrate upon the true independent thinker and see just how his or her mind works.

Dr. Velikovsky's Comet

Of all the independent thinkers of modern times, none had caused more controversy than Dr. Immanuel Velikovsky. Though he is of Russian birth (he was born at Vitebsk, in 1898) he has lived for many years in the United States.

Briefly, his theory is that the planet Venus used to be a comet—and that in Biblical times it passed close to the Earth more than once, causing tremendous global upheavals, and behaving rather in the manner of a cosmical ping-pong ball.

This in itself is extremely revolutionary, and Dr. Velikovsky's original book, *Worlds in Collision,* caused raised eyebrows when it first burst upon the scientific world in 1950. Like Dr. William Whiston, a contemporary of Sir Isaac Newton, Dr. Velikovsky believed that comets can change into planets, and vice versa.

Let it be said at once that Velikovsky's doctorate is perfectly conventional. For some time he practiced as a GP in Jerusalem. He then trained as a psychoanalyst, and practiced at Haifa and Tel Aviv. It was not until 1939 that he went to America. He is not a trained astronomer or mathematician, but psychoanalysts and psychiatrists in general are not to be classed with other people.

All of his three books are heavily annotated, and every Biblical reference is correct. In fact, Dr. Velikovsky has done his homework extremely well in this respect. It is his scientific reasoning that has caused his contemporaries to frown thoughtfully or, in some cases, to explode in anger.

He begins fittingly enough, with the giant planet Jupiter, which—he says—suffered a tremendous outburst, and shot out a comet that later became the planet Venus. Originally the comet Venus had an elliptical path, and this led it into a series of strange encounters. It first passed near the Earth in 1500 BC, at the time when the Israelite exodus was being led by Moses. The result was that the Earth temporarily stopped spinning, or, at least, slowed down; and the Red Sea was left high and dry for long enough to allow the Israelites to cross. Then, however, came huge upheavals, as the Earth's surface twisted and turned under the influence of the gravitational pull of the comet Venus. Petrol rained down—and, incidentally, our modern motor cars are powered by precisely this petrol, "remnants of the intruding star that poured forth fire and sticky vapor."

Conveniently, the Earth's rotation started up again just in time to swallow up the Egyptians who were in hot pursuit on the retreating Moses. But this was not all. Having made its presence felt, the comet Venus withdrew into limbo, but came back for a second visit two months later so that it could produce the lightning, thunder, and other spectacular effects noted when Moses was given the Ten Commandments on Mount Sinai. Subsequently, some of the materials in the comet's tail rained down on Earth in the form of manna upon which the Israelites fed for the next forty years.

One might have though that this would have been enough. But no! The comet Venus continued its erratic career, coming back to see us several times—once, for instance, to shake down the walls of Jericho. However, the day of reckoning was at hand. The comet collided with Mars and had its tail chopped off, so that it stopped being a comet and started being a planet. Not to be outdone, Mars itself moved closer to the Earth, and indeed nearly scored a bull's-eye in the year 687 BC. Further encounters took place, and Dr. Velikovsky links various Old Testament disasters with those approaches of Mars. Then, gradually, the situation eased. Mars retreated again, and resumed its former path; Venus, no longer a comet, settled down into a peaceful and almost circular orbit; the earthquakes, tidal waves, floods, cyclones, and other terrestrial events subsided.

Patrick Moore
Can You Speak Venusian?
A Guide to the Independent Thinkers
Norton, New York, 1972

Chemiker Anekdoten

Joseph Hausen compiled a 91-page book in which he gave glimpses into the lighter side of the lives of 66 famous chemists. These excerpts from that collection were translated by Julian Smith

Walter Nernst: Professor Nernst in retirement raised carp. When asked why not chickens, he said he preferred animals in thermodynamic equilibrium with their environment; why spend money to heat the world's empty space?

R. B. Woodward: At the Munich (1955) meeting of the Gesellschaft deutscher Chemiker, Woodward attracted attention as he roamed the halls carrying a big notebook in a blue silk cover on which was embroidered the structural formula of strychnine. The next day he appeared bearing a cover innocent of any embroidery. Asked a friend, "Why no structural formula?" Quipped Woodward, "Oh, I'm traveling incognito today."

R. W. Wood: In Paris Professor Wood ate in a small boarding house. When there was fowl for dinner he astonished his tablemates by sprinkling a white powder on the bones left on the plates. Next day he brought along a small alcohol lamp and tested a drop of soup in the flame. Seeing a red flash, he chortled triumphantly, "I knew it!" Then he explained to his astonished audience, "I only wanted to know if the bones go into the soup, so yesterday I sprinkled them with lithium chloride." The tell-tale lithium red flame test was confirmation.

Joseph Hausen
Journal of Chemical Education
Vol. 35, pp. 237–466, 1958

Dimitri I. Mendeleev (1834–1907)

The best-known Russian chemist developed the periodic classification of the chemical elements independently of, but at about the same time as, Lothar Meyer (1830–1895). His prediction of the properties of some of the then unknown elements was masterly. He made extensive studies of the properties of gases and liquids and anticipated Andrew Andrews (1813–1885) with regard to the critical temperature. Mendeleev thoroughly investigated Russian petroleum. His two-volume textbook in both the original Russian and in translations had much influence.

Mendeleev tried to bring into inorganic chemistry some of the same order that was then emerging in the sister field of organic chemistry. He prepared a set of large individual cards for the elements, and on these wrote the atomic weight, the physical and chemical properties, and the corresponding data for the compounds of the particular element. He then began a lengthy game of solitaire, trying different ways of arranging the cards. He took the problem to bed with him, and according to his own account, there came to him in his sleep the sudden inspiration that arranging the cards in the order of their increasing atomic weights yielded a periodicity in the properties of the elements. This discovery made him one of the most talked of chemists in Europe. Similar fame came to Lothar Meyer, the German chemist, who had arrived at a comparable arrangement at about the same time, but on largely physical rather than chemical grounds.

Some years later, it was announced that Mendeleev would attend the national meeting of the British Association of Science, and a large crowd assembled. When the assigned place was reached in the proceedings, a bearded gentleman, who obviously was a foreigner, arose to thunderous applause. When order was restored, the guest announced to the great disappointment of all, "I am not Mendeleev. He has asked me to express his apologies to you but he does not speak English. My name is Lothar Meyer." A moment of silence and then great applause. Lothar Meyer then motioned to somebody behind him; another hirsute foreigner came forward, whom the audience immediately recognized as Mendeleev. The resulting tumult was never forgotten by those fortunate to have witnessed this historic occasion.

Mendeleev's first marriage came about mostly because his sister felt he needed somebody to look after him. The couple had two children, but he and his wife quarreled almost constantly about how the children should be trained. Finally they lived apart, he occupying their town house while she lived in the country. In 1876, when Mendeleev was more than forty, he fell violently in love with an art student who was only seventeen. He was a national figure, yet he pursued her ardently; her parents finally sent her to Rome to continue her education and quiet the notoriety. Mendeleev hurried to Rome, leaving a note behind to the effect that he would throw himself into the sea if he could not have her. But she assured the great scientist that he had all of her love and she would marry him as soon as he could secure a divorce. Mendeleev returned to Russia; a satisfactory financial settlement was arranged with respect to his wife and the children, and the divorce was granted. Then he learned that Russian law, both civil and ecclesiastic, forbade the remarriage of a divorced person until seven years had elapsed. This legal hurdle did not fit into his plans at all. After some search, Mendeleev found an Orthodox priest who was willing to defy the civil and church authorities, for a price. Two days after the wedding the secret came into the open. The priest was defrocked and departed secretly, but richer by ten thousand rubles. The civil authorities took no action against Mendeleev, the noted chemist, despite that fact that officially he was a bigamist.

A divorced nobleman, who wished to remarry before the seven years were up, appealed to the Czar and cited Mendeleev in support of his petition. The Czar replied: "I admit that Mendeleev has two wives, but I have only one Mendeleev."

Mendeleev's second marriage was a real success. His wife introduced him to a whole world of art and related fields; in short, she made him a more cultured person. They had four fine children. When the Bolsheviks came into power after World War I, they remembered that Mendeleev had always been on the side of the people and had opposed many of the Czarist political ideas. The Allies supported the anti-Communist government and blockaded the Soviet ports. Food became scarce and had to be rationed. However, the Leningrad authorities, remembering Mendeleev's efforts in behalf of the people, issued a double set of ration cards to his widow. As a result, Madame Mendeleev became easily recognizable because she was the only fat woman in this great city, now Leningrad, formerly St. Petersburg.

In 1906, just a year before his death, and more importantly, almost forty years after the announcement of his great achievement, an effort was made to have the Nobel Prize awarded to Mendeleev. However, when the votes were counted, he lost by just one vote to Moissan. This surprising result was explained by a close contemporary: "For older works they should be considered only if their significance did not become apparent until recently." Had he done his great work after the establishment of the Nobel Prizes (Nobel died in 1896 and the first prizes were awarded in 1901), he doubtless would have been chosen, but by 1906 it was too late to honor Mendeleev for work done long before, at least in the opinion of a tiny majority of those entitled to vote.

Ralph E. Oesper
The Human Side of Scientists
University of Cincinnati Publications, 1975

Dr. Watson's War Wounds

Dr. J. H. Watson, who died twenty years ago at the age of 103, was a man deserving of more honor than he received. I would go so far as to assert that he was in fact the brain behind the exploits of Sherlock Holmes, and that only his innate modesty prevented his true worth from being known; but this is a matter that must be discussed elsewhere. The problem I wish to investigate here is a surgical one: the situation of Watson's war wounds.

Let us examine his own account of the action of the disastrous field of Maiwand in Afghanistan. He tells us in *A Study in Scarlet:* "There I was struck on the shoulder by a Jezail bullet, which shattered the bone and grazed the subclavian artery." As everyone knows, he was saved by his orderly and brought to the base hospital in Peshawar where he contracted typhoid fever; on his recovery he was invalided home. But later, in *The Sign of Four,* he tells us: "I made no remark, but sat nursing my wounded leg. I had a Jezail bullet through it some time before, and although it did not prevent me from walking, it ached wearily at any change in the weather." And in still another place Holmes describes Watson as a half-pay officer with a damaged tendon Achilles. This last claim can, I think, be summarily dismissed. A bullet, even of the Jezail variety, that passed through Watson's leg could scarcely have damaged his Achilles tendon; he is obviously referring to a soft-tissue wound, and he says his leg, not his ankle. Still less would a bullet in the shoulder have caused his leg to ache. The truth of the matter is surely that Holmes did not know where the Achilles tendon was; or perhaps that he was thinking of Achilles himself, who by a strange coincidence was shot in this very situation, though not with a Jezail bullet, by Paris and Phoebus Apollo in the Skaian Gate. We have Watson's testimony that Holmes was ignorant of Carlyle; is it not conceivable that one so ill-informed might have confused his friend Watson with another military man, Achilles? But when Watson's own testimony seems to contradict itself we must consider the facts more carefully.

The facts are simple. Watson tells us in one place that he was struck on the shoulder, and in a later account that a bullet passed through his leg. It is quite clear from his first account that there was only one bullet on the field of Maiwand. Now Watson was a London MD, and ought thus to know the difference between the shoulder and the leg; we might expect such knowledge of even a medical student. I think we must therefore accept both his statements as true. The questions then arise; How did he come by two wounds with one bullet, of (if there was more than one bullet) could he have been wounded in the leg, also with a Jezail bullet, on some later occasion?

Let us deal with the second question first. Baring-Gould[1] has made the extraordinary assertion that Watson may have received this second wound while engaged in the well-known affair of the Vatican cameos. This is most improbable. The

[1]Baring-Gould, W. S., *Sherlock Holmes*, London, 1962.

dangers that an Englishman of the period might incur in visiting the Vatican were doubtless great, but Jezail bullets were not among them. It seems equally improbable that one of the Afghans who had fought at Maiwand should have taken the trouble to journey all the way to England merely in order to shoot a half-pay officer in the leg with his primitive matchlock. Such unrelenting malignity, not to say inefficiency, is scarcely credible. There is a third possibility; however. Could Watson not have brought back an Afghan musket as a memento of the campaign, and not realizing that it was loaded have shot himself with it in England? Or could Holmes have shot him? We know that Holmes was uninhibited in his use of firearms; he would expend a hundred Boxer cartridges in decorating the wall of his sitingroom with a patriotic VR in bullet hules, but there is nothing in the sacred canon to suggest that either he or Watson was ignorant of the precautions a sensible man ought to take. They were in fact, both marksmen of the caliber that could hit a running hound at dead of night in a thick fog with a short-barrelled revolver, a job for which most people could have taken along a shotgun and still expected to miss. Besides, there is no mention among the well-remembered furnishings of Baker Street—the gasogene, the Turkish slipper, the coal-scuttle that contained the cigars—of an Afghan muzzle-loader adorning the bullet-scarred wall or sharing a corner with the fifty-five shilling Stradivarius. No; Watson brought back no musket from Afghanistan. How then did he come to be injured in both the leg and the shoulder on the field of Maiwand? And why did he mention only one wound in first account?

Let us go back to his accounts. He tells us that a Jezail bullet passed through his leg, and that a Jezail bullet struck him on the shoulder.   When every other possibility has been excluded the one that remains, however unlikely, must be the true explanation. There was only one bullet; therefore it passed through Watson's leg and lodged in his shoulder. The only difficulty that remains is to determine the position he was in at the time of this unfortunate occurrence, and why he omitted to mention it in his account of the battle.

I have given this matter a great deal of thought, and until recently could come to no conclusion. There has lately come into my possession, however, part of a manuscript that I have no hesitation in accepting as genuine. It purports to be Watson's diary. Much is illegible; some is missing; and parts of it, such as the altogether fascinating account of the politician, the lighthousekeeper, and the trained cormorant, deal with matters that certainly cannot yet be made public. But it does tell how Watson got his wound. Briefly, he was squatting over the edge of a precipice, of which there are many in Afghanistan, to answer the call of nature; and he was fired on from below, the bullet passed through the adductor muscles of the left thigh and struck him on his shoulder with the effects he has described.

It has been objected that the account cannot be genuine because no marksman would take up by choice so hazardous a position, but this is surely to underrate the intrepidity and agility of the hardy Afghan mountaineer. It is very easy to criticize, but if anyone has come up with a better explanation than this I have yet to hear of it.

We see now why Watson referred only to his shoulder wound in his first account. The reader will remember that the more prudish of the Victorians scrupled to refer even to the legs in conversation, regarding them as vaguely indecent organs. Watson, in his frank and manly way, would have had none of this. But he was a child of his age nevertheless, and although the episode was still green in his memory he could never bring himself to explain exactly what he had been doing on that inauspicious occasion. After a decent interval he mentioned the wound in his leg without explaining it. For Watson was the most modest of men. I am sure that Holmes never knew what had happened.

The modesty of Watson's, of course, explains why he let Holmes get away with all the credit for his cases, and indeed played down his own achievements and contributions in the self-deprecatory way so characteristic of the English gentleman. Besides, there was the General Medical Council, which frowned on anything that might resemble advertising by a doctor. It was distasteful for Watson to blow his own trumpet, and it might get him into trouble as well. But the events in which he was involved were exciting, and he was a writer; so he blew Holmes's trumpet instead. And Holmes, the poor deluded creature, believed every word of it. Watson was a genius; a good doctor, a most graphic writer, a brilliant detective, and an honorable gentleman, even if he did have that spot of trouble in Afghanistan. It is time that his merits were recognized.

Peter Brain
The Lancet
Vol. 2, p. 1354, 1969

Paul Flory

Flory is well known for his careful research and profound reasoning in many branches of polymer science, and his book, *Principles of Polymer Chemistry,* is a classic.   His many honors include the Priestley Medal and the Nobel Prize in chemistry (1974).

Flory had left the academic field fairly early for a career in industrial research at duPont, Standard Oil Company of New Jersey, and Goodyear Tire and Rubber Company, but his experiences finally convinced him that industrial research was not his proper metier.

In those days, Flory was not known for his diplomacy or patience in dealing with lesser intellects, and he could not get on at all with Goodyear's director of research.   The director promulgated many fussy restrictions on working hours, opening windows, closing drapes, and so on, which Flory considered intolerable pettifogging and encouraged his group to ignore as a matter of principle.   He left Goodyear in 1948 to return to more academic environments (Cornell, Mellon Institute, Stanford), where his highly fundamental and theoretical approach would be better appreciated.   Asked why he left industry, Flory replied: "I got tired of casting synthetic pearls before real swine."

Karl Ziegler

Even some eminent representatives of the German chemical industry did not escape the occupational hazard of the industrial scientist: looking upon his academic colleague as an intellectual dilettante whose ideas and work are of dubious practical import. Bayer, one of the big German chemical companies, never took a license from Ziegler, although they had an early opportunity. Part of the reason may have been the attitude exemplified by Bayer's research director, Dr. Otto Bayer. He was among a number of leading German chemists present at a dinner that Ziegler attended and at which he was asked about the importance of the new chemistry coming out of the Max Planck Institute. Ziegler, with characteristically dignified assurance, said he was sure that it would stand as an important contribution and would be known in the future as a "Mulheimer Chemie" (after the town Mulheim). Bayer, polishing his reputation for sarcastic wit, remarked that that would be an unfortunate choice of names for an internationally famous process. The problem would be that all Frenchmen would be sure to mispronounce it as "Mull-eimer" (the initial "h" being silent in French). In German, "Mulleimer" means "garbage can"!

Ziegler was not amused. Nor was he deterred: when several years later he published his first major paper on polyethylene, it was titled, boldly, "Das Mulheimer Normaldruck Polyathylen-Verfahren" (The Mulheim Normal-Pressure Polyethylene Process). However, it was but a short while after that until someone else (Giulio Natta) rechristened it and put the phrases "Ziegler Chemistry" and "Ziegler Polyethylene" into the language and the literature.

Frank M. McMillan
The Chain Straighteners
Macmillan, p. 40, 1981

Sir Alexander Todd Recalls

Our work on *Cannabis* at the Lister brought me into an early and, in retrospect, slightly absurd confrontation with the Home Office Drugs Branch. The starting material for our studies was a distilled extract of hashish that had been seized by police in India and had been obtained from them by my colleague Franz Bergel while on a visit to that country some years before and while he was still resident in Germany. The distilled resin was transmitted to Germany, via the diplomatic bag, and, in due course, brought to Edinburgh through the port of Leith together with a variety of other chemicals in a suitcase carried by Bergel; no questions were asked by the Customs. In starting our work in the Lister we first isolated cannabinol from the resin, and showed that, contrary to general belief, it was pharmacologically inert, the hashish effect residing in the material left after its removal. We submitted a brief paper on these observations to a meeting of the Biochemical Society early in 1938 and this was duly printed in *Chemistry and Industry,* which in those days published short abstracts of papers read at such meetings. Within two or three days of the appearance of our little note I received a letter from the Drugs Branch inviting me to come to the Home Office and speak with one Inspector X at my early convenience. This interest of the authorities in me and my work was unexpected, but I went to the Home Office and was duly shown to the room of Inspector X, who was seated at a large desk on which lay a copy of *Chemistry and Industry* and what looked like a large ledger. After exchanging the usual courtesies the Inspector said, "I see you have been doing some work with hashish," to which I could only reply "Yes."

"You realize, of course, that *Cannabis* in all its forms is proscribed."

"I suppose it is."

"Well, we can probably straighten things out fairly easily so don't worry. Here in this book I have a record of all the legal holders of *Cannabis* in this country with the amounts of material they hold."

I glanced quickly at the opened ledger. On the page visible to me there were a dozen or so names of doctors and professors each with small amounts of drug opposite them— usually only a few grams or ounces.

"Now," said Inspector X, "presumably you got your hashish from one of these holders and the only irregularity is that he didn't notify me; but we can easily put that right by an appropriate entry. Which of them gave it to you?"

"I'm afraid none of them did."

"Then who did?"

"The Indian police."

"Yes, but how did it get into this country?"

"In a suitcase at Leith."

"You mean that you smuggled hashish?"

"I wouldn't call it smuggling. It was in a properly labeled flask, but the Customs people didn't seem to be very interested in it; it was just one of a number of bottles of chemical specimens in the suitcase."

At this point there was a brief silence; then, "How much of the stuff have you got?"

I confess I had been waiting, not without trepidation, for this question and at first I tried to parry.

"Well, of course, what I have is a distilled extract of hashish and not the drug as it appears on the Indian market."

"Never mind about that—just tell me how much."

I plucked up my courage. "Two and a half kilograms."

"Good God!"

The Inspector looked worried and after a few moments he said "What are we going to do about this?" followed by a long pause and then "I think we had better make you a licensed holder of *Cannabis*."

So he wrote in his ledger "Dr. Todd 2-1/2 kilos," and added, "You will of course understand that this material must be kept under lock and key and all amounts you use in your work must be duly recorded, and that your records will be open to inspection by us at any time. Furthermore, if you publish any papers arising from work with this resin we will expect twenty-five reprints of each paper."

"Certainly. Where shall I send the reprints?"

"Send them to me at the Bureau of Drugs and Indecent Publications."

*   *   *

However, my chemical defense commitments had their lighter moments. I recall being called upon to travel down to the Defense Research Establishment at Porton to watch a demonstration of a new chemical weapon for use against tanks. It must have been in 1941, because air raids were heavy and frequent, tobacco was very scarce, and, since petrol was equally hard to come by, I traveled down from Manchester by train through Bristol to Salisbury. As it happened, Bristol had a big raid on the night my train was passing through and we had to lie stationary in the railway yards with bombs dropping uncomfortably close until the raid ended. We then trundled on through the placid countryside of southwest England and arrived at Salisbury around 8 AM. where I was to breakfast before setting out in an army car for the demonstration on the open plain near Porton. Now, in those days I was quite a heavy cigaret smoker and the modest supply I had wheedled from my supplier in Manchester had long since gone and I was rather desperate but, needless to say, I could find no one in Salisbury who would supply me with any. So I breakfasted, trundled off to Porton, watched gloomily a rather unconvincing weapon demonstration, and was taken to the local officers' mess for lunch. After a wash I proceeded to the bar where— believe it or not—there was a white-coated barman who was not only serving drinks but also cigarets. I hastened forward and rather timedly said "Can I have some cigarets?"

"What's your rank?" was the slightly unexpected reply.

"I am afraid I haven't got one," I answered.

"Nonsense—everyone who comes here has a rank."

"I'm sorry but I just don't have one."

"Now that puts me in a spot," said the barman, "for orders about cigarets in this camp are clear—twenty for officers and ten for other ranks. Tell me what exactly are you?"

Now I really wanted those cigarets so I drew myself up and said "I am the Professor of Chemistry at Manchester University."

The barman contemplated me for about thirty seconds and then said "I'll give you five."

Since that day I have had few illusions about the importance of professors!

*   *   *

While in Chicago, and indeed for some time before that, I had been subject to rather tiresome bouts of indigestion, and early in 1949 matters came to a head rather suddenly, and I had to undergo major surgery for removal of my gall bladder, which was completely blocked by a large lump of beautifully crystalline cholesterol. The Dyestuffs Division Research Panel of Imperial Chemical Industries Ltd. used to meet on the first Friday in each month at Blackley Works, Manchester, and its three external members—Robert Robinson, Ian Heilbron, and myself—used to travel from London by rail on the evening before and stop over at the Midland Hotel in Manchester. When I had recovered from the gall bladder operation sufficiently to go up again to a Panel meeting, I found myself traveling up on the evening train with Robert Robinson. Robinson, my former teacher and now a close friend, had, as one of his more endearing characteristics, a habit of reacting emotionally and almost violently at times to comments on chemical matters that were brought suddenly to his notice. At the time of which I am writing, he was in the throes of his large-scale effort to synthesize cholesterol—a rather fashionable pursuit that then occupied a number of competing groups throughout the world. Robert proceeded to recount to me that evening a particularly nasty snag that was blocking his path. After we had discussed it for a bit I said "I'm sorry you are having difficulties—I don't suppose you know it, but I have just completed a synthesis and isolated cholesterol absolutely correct in structure and stereochemical configuration." He rose to the bait like a trout to the mayfly, and almost shouted "What do you mean? Why was I not informed that you were working on cholesterol? How long has this been going on?" I replied "Don't worry, Robert, it's just a little something I did entirely on my own in my spare time"—and with that produced my beautiful gallstone, which I was carrying in my waistcoat pocket. Robert's wrath vanished as quickly as it had come, and was replaced with a roar of laughter. When we got to Blackley next day, I heard him recounting my successful syntheses to all and sundry.

Sir Alexander Todd
A Time To Remember:
The Autobiography of a Chemist
Cambridge University Press, 1983

Geologists

Those who initiate memorandums could be doing more harm than they are, and might actively be participating in teaching and research and meeting with students.

The worth of an academic program is related inversely to the thickness of the file on it. If the program fails completely, there are certain to be enough memorandums and reports for a book.

*   *   *

A committee is appointed for problems that cannot be solved, problems that no one wishes solved, problems that need not be solved, problems so tiny that a secretary will not be bothered with them, or to divert suspicion and pressure away from the instigator and take him or her off the hook.

*   *   *

If you can't face a class, schedule movies. If there are not enough movies, show slides of your equipment. If your slides are insufficient, arrange for a colleague to present several lectures. If your colleagues are reluctant or too few, tell your teaching assistant to prepare to take the class for several periods for the experience. In the time that remains, give three tests.

*   *   *

To be a good human being is no longer sufficient. You must be the best, a champion, a winner. The game is winning, and all important activity is reduced to the game. One cannot be happy unless one is successful, and one is not successful unless one wins. However, survival as a human being is unrelated to these rules, and nearly impossible within them.

Field Chips

Your own field problem is ruggeder than anyone elses, and requires skill and courage not found among most field geologists.

*   *   *

The best field geologist is the one that has seen the most rocks, but only one look is permitted at each outcrop. If you must return several times, it indicates that you are inefficient, unobservant, and unable to grasp complex relations in a short visit. Such plodders may be happy in sedimentary petrology.

*   *   *

Some administrators believe that any geologist can do a field problem and little training or skill is required. A gifted geologist therefore is urged not to waste time in the field and to solve problems in the laboratory. Rocks from the departmental collection can be used for laboratory studies, or the geologist can collect new ones on a weekend while out for a bit of air with the family.

*   *   *

It requires a lot of beer to do field geology; the salt is needed to keep going.

*   *   *

No mapping project is as tough as the last one. Sorry that you weren't along on that one.

Exploring Anthills

I was interested in sand before I was interested in anything else in geology. Sand and ants.

When I was five I began a study of the anthills on the terraces above the Big Horn River in northwestern Wyoming. Almost immediately I found two types of ants, red ones and black ones. They differed in size, and the big red ones were the mean babies. That was about as far as I got with ants. Although I occasionally covered a few of them with sand to test their burrowing capabilities, which are formidable, I generally left them alone. I even fed them now and then, mainly to watch their carrying ability. All in all, I believe they may have appreciated my actions; at least I wasn't bitten often.

I admit I didn't make many sharp observations about ants. And I didn't become a biologist either. My excuse is that I was more interested in sand than ants, a propensity that continues.

At age five, and for the next several years as I continued my study, the sand was almost a world in itself. Red grains were rubies or garnets. Green grains were surely emeralds. And I hoped to find a diamond that would make me and my family rich. I strained my eyes to find one, and scratched dozens of windows to test hundreds of possibilities. But it didn't happen.

I put in a lot of time on those friendly anthills looking at sand grains. More than a gunny sack full of sand had accumulated before my interests changed and I stopped spending so much time with the anthills. However, even today I drop down on my stomach now and then to renew old friendships.

M. D. Picard
Grit and Clay
Elsevier, Amsterdam, 1975

Huxley in Repartee

In the *Origin of the Species*, Darwin has said virtually nothing about humanity except for that almost final comment: "Light will be thrown on the origin of man and his history" (1859, p. 488). But no one was fooled by the brevity of this reference, and almost every attack on Darwin brought in the "monkey question." Yet, though it was religion that made the debate over humanity so intense, there was a factual level—as close to pure science as anything. Here we see the themes of science, philosophy, and religion intertwining.

The question at issue was whether humans are in any essential way different from other animals, particularly the great apes. That we are was the position of Richard Owen, who, with all his great authority, had asserted that the human brain "presents an ascensive step in development" because, among other things, it alone has 'the hippocampus minor,' which characterizes the hind lobe of each hemisphere." This claim probably led to the most famous of the clashes between the Darwinians and their opponents, that between Huxley and Bishop Wilberforce at the British Association meetings at

Oxford in 1860. At a midweek meeting, Owen restated his claim about the differences between man and other animals, and Huxley contradicted him flatly... Infuriated, Owen primed Wilberforce, who returned to the attack on Darwinism at a Saturday meeting. Apparently somewhat charmed by his own rhetoric, Wilberforce asked Huxley (also on the program) whether it was through his grandfather or his grandmother that he claimed descent from monkeys. Trading comments of this sort with Huxley was not safe, as the bishop soon found out. To the delight of the Darwinians the story got around that Huxley had retorted that he would rather be descended from a monkey than from a bishop of the Church of England. More probably, but hardly less scathingly, he replied that if faced with the question, "Would I rather have a miserable ape for a grandfather or a man highly endowed by nature and possessed of great means and influence, and yet who employs these faculties and that influence for the mere purpose of introducing ridicule into a grave scientific discussion—I unhesitatingly affirm my preference for the ape."

Michael Ruse
The Darwinian Revolution
University of Chicago Press, 1979

Taking a Stroll Down Babel Street

There are many shops in Babel Street. They range from Fred Fortran's successful grocery chain to Dr. Avery Tower's latest toy shop. To gain a better understanding of the business, Peter Brown talked to some of the proprietors, beginning with the pernickety grocer himself.

Of all the grocer's shops in town, Fred Fortran's does the most business. In fact only one shop of any nature does better and that is Cobol's stationers. Fred has been around for untold years and the doctors expected him to die long ago. Yet he continues working, looking the same as ever, watching his rivals come and go.

He is a man of very precise habits and it was important that I arrived at the right time to interview him. His day is broken into eighty periods, and various parts are set aside for different activities. This is one of his rules. If you arrive in the wrong period he might refuse to see you, or worse, take you to be someone else.

Fortunately I was all right and walked straight into his office above one of his shops. He greeted me cheerfully, but did not interrupt his work.

I started by asking him why his shops sold so few lines, and why they had remained unchanged for years.

"I won't stock any of these fancy new-fangled foods," he replied, "what was good enough when I was a lad is good enough now. My customers have got used to the foods I stock and it would be madness to change."

I had brought him a box of assorted subscripts as a present, but, sensing that he might not be pleased with it, I decided to explore the ground first.

"I see you are still stocking the old Brand 704 or Brand 709 plain subscripts."

"Yes, sold well for many years, those have," he said proudly "no-one else makes them like that any more."

I remembered that most items in his shops were labeled Brand 704 or Brand 709. Obviously my intended present, a relatively new brand, would not be welcome so I decided to forget it.

I moved on to an even more controversial issue, and told him that many people accused him of spreading the debilating disease of *gotoitis*, which results in a considerable loss of productivity.

"Rubbish!" he said, "Everyone has gotoitis but it doesn't do them any harm. Nowadays some people want to cover it up by wearing fancy clothes, but it's still there underneath, you know. Back in the good old days . . . ."

He continued talking of the good old days for some time. When at last he finished I asked him about some details of the running of his shops.

"Let's say I ask you to repeat an order X times," I said, "and I tell you X is zero. Why do you execute the order once?"

"Your question is phrased badly, I won't accept it," he said crustily. When you say X I assume you don't mean an integer. It's one of my rules. And if you don't mean an integer, I won't execute your repeat instruction at all. What's more I would reject all your other instructions until the error is corrected."

"All right, make it N times," I said, amazed at the escalating consequences of my original slip.

"That's better. I will accept N times. To answer your original question, if you tell me to do something I get on with it. I don't waste time asking if you really didn't want it done after all. When I have finished the first time, I look to see how many more times I have to repeat the order. I try to be efficient, that's the point—not like Lady Algol, who spends so much time deciding whether or not to do anything that every job takes twice as long."

He continued his deprecating remarks about Lady Algol, who owned a rival chain of shops.

"...and the most ridiculous thing of all is that there is no standard input/output system in her shops."

Eager to change the topic of conversation I asked him about his own input/output.

"Oh, mine is very simple," he explained, "you have formats with labels attached and within the formats . . . "

He tried hard to explain it to me, but I never did manage to understand.

We then talked about his worldwide success, a shop in every town. His plan has been to make every shop the same, so that customers could adapt easily from one to another. Every detail of the shops has been specified by Miss Ansi, who is very particular about such things. Fred was a bit resentful of her, however.

"She is very keen to standardize other people," he complained, "but with herself it's different. She even keeps changing her own name."

The standardization shops had been quite successful ("Compare it with the mess the Algol shops are in," said Fred), but the size of some goods varied a bit and many shops had introduced some new lines of their own. I had even bought my box of assorted subscripts in one of Mr. Fortran's other shops.

Throughout our interview, Fred had continued to work away at his accounts. It amazed me the work he got through. He did not show any imagination, just thoroughness and efficiency, and the result was low prices and lots of customers.

It was time to leave. I thanked him and gathered up my things. As I left I asked him where the toilet was.

"You go to Room 163."

"Doesn't it have a name on the door?"

"No," he said, "I only allow people to go to rooms with numbers. It's one of my rules."

Peter Brown
Computing Europe
October 31, p. 12, 1974

Meyer's Itch

Someone has said that the way to be remembered is to name a disease after yourself. First, however, you must find a disease that has not already been named. I am happy to report that I have found such a disease: Meyer's itch.

Since humans can be attacked by, and acquire a dermatitis from, the schistosome cercariae of birds, it is only reasonable to expect that the reverse also occurs: Birds are susceptible to invasion and dermatitis from human schistososme cercariae. Meyer's itch, then, is a dermatitis of the extremities of wading birds, that have been attacked by the cercariae of human schistosomes. You may well have seen a case of Meyer's itch if you have ever observed a wading bird standing on one leg, which it is rubbing madly with the other. While this disease may not be of much importance for people, I can assure you it is for the birds.

Ernest A. Meyer
Microorganisms and Human Disease
Appleton-Century-Crofts
New York, p. 415, 1974

The Faded Majesty of Lady Algol

Peter Brown continues his exploration of Babel Street with a visit to the aging Lady Algol who, despite her great fecundity, still retains a measure of the beauty that won her much success abroad.

Lady Algol has a shop at number 60. A rival shop, run by a relative, is at number 68, but has not, as yet, had much effect on her business.

The shop impressed me, but when I went inside I was even more impressed by Lady Algol herself. Although she is getting old now, I was still struck by her grace and beauty. One could well understand how she appealed to the youth of her day. Yet there were signs of old age—a few wrinkles and a very slightly faded look about her. Perhaps this is why her appeal is not so strong to today's youth.

We talked of the past. She reminisced enthusiastically of the times when every eager academic in Europe sought after her.

"Did you also visit America?" I asked her.

"Well, yes," she said, with rather less enthusiasm, "but I did not make quite such a hit over there. I wanted to set up some of my shops, but the Americans thought my overheads were too great and my prices too high. Also in those days people were not even sure how to build my shops at all. Nevertheless I think that the main reason for my relative lack of success in America was that they resented that many of my roots are in Europe. They prefer All-American girls."

"Were they right in thinking your shops aren't efficient?" I asked.

"It depends on the premises. Nice Mr. Burroughs has provided some excellent premises, designed specially to meet my needs. He is one of the few Americans who really love me, and he has remained faithful for years. He buys almost everything from my shops, but has made me introduce a lot of new lines. Some of my friends think he has gone too far, and has destroyed the nature of my shops, but I don't think so."

"I see that your prices vary from shop to shop," I continued, "but how do they compare on average with those at Fred Fortran's stores?"

"Don't talk to me of that common and cheap Mr. Fortran," she said severely, "I refuse to be compared with him."

I should have realized what a faux-pas it was to mention such a name in the majestic atmosphere of one of her shops. (I smiled to myself as I remembered Fred Fortran's comments about Lady Algol.)

"Sorry, My Lady," I said, obsequiously, and quickly changed the subject.

We talked of her children, of whom she was very proud.

"How many do you have," I asked.

"Hundreds! They are all over the world, and most of them have different fathers."

The number seemed a bit excessive to me. Perhaps she was counting her grandchildren and sister's children, as well. However there was no denying that there were many fruits of her boisterous youth. Moreover, the geneticists regarded her a good material for breeding. They had even paired her off with Mr. Cobol, but she talked without warmth of their relationship.

"It was an arranged meeting, and I don't think we ever had any real feeling for each other. And he does talk so much!"

Mister Cobol must have done more than talk, because they produced an offspring, PL/I, one of whose shops I was planning to visit later. I think Lady Algol gained some spiteful pleasure out of the result because Mr. Cobol's arranged marriage with one of Fred Fortran's daughters failed to produce any issue.

I turned the conversation round to the goods in her shops. All the shops sell much the same basic lines, but the design of the shops varies greatly and her goods always seem to be packaged in different ways. This is a great nuisance if you move from one shop to another. I wondered why she did not follow Fred Fortran's example, and adopt some standardization (Though even he wasn't always successful in this respect). However I did not dare mention Mr. Fortran's name explicitly again.

"Why do your shops vary so much?" I inquired.

She immediately became embarrassed, and talked round the subject for a while, saying that it didn't really matter. Finally, after I had asked some more pressing questions, she broke down and told me the full story.

When her company was started, the board of directors consisted of thirteen people, covering seven different countries. To many people's surprise the board did a good job. Contrary to the usual practice of multinational committees they agreed on something significant and almost free from ambiguity. However they made some omissions—intentional at the time— the most significant of which was a plan for how to get in and out of the shop. The result was that every shop manager designed his own input/output system. No manager was going to be seen to copy anyone else's system, so all the shops developed differently. (Many of the original shops were designed by Dr. Avery Tower and his followers. By the very nature of their designers these shops were all unique.)

Thus anarchy reigned, and once the shops had become established in this way it was too late for the board to do anything about it. In any case the board didn't meet very often since they had mostly gone on to other things.

I changed the topic of conversation to happier matters and asked about some of her famous customers. Her favorite was Mr. A. C. M. Algorithms, who had made regular monthly visits for many years, though recently he had taken to frequenting other shops as well.

Before I left her she showed me round the shop. It was all very orderly, with large units divided into smaller units and these in turn subdivided into smaller units still. I only noticed a few blemishes; an *else* dangling from a shelf and her chocolate *declares*, own brand, which looked a terrible mess.

She showed me some of her famous tricks. Two mirrors rearranged in the shop so that you can see an image of herself within herself within herself, and so forth. This had acted as a great enticement during her youth, and it still remained seductive. She also showed me Jensen's machine for picking goods off the shelves in a variable way. I must say I didn't understand it properly, and it appeared difficult and expensive to operate. In my ignorance I thought it would be easier to take things off the shelf without the aid of the device.

As we walked around the shop at the end of our interview, I noticed more signs of her age. She was not quite as steady on her feet as she used to be. Would she still be flourishing in ten year's time! Maybe she would not. But one thing was certain: through her children and grandchildren she would remain a major influence worldwide.

Peter Brown
Computing Europe
Nov. 6, p. 10, 1974

Auklet Double Dactyls

Some liberal (sloppy) *Auklet* editors have accepted attempts at double dactyls (and limericks, too) that do not follow the rules. All those included here are *echt* double dactyls.

A double dactyl is a poem in two stanzas of four lines each. Lines 1, 2, and 3 of each stanza consist of six syllables, stressed on the 1st and 4th, as in the phrase "ART-i-choke MAR-ma-lade." Line 4 of each stanza omits the last two syllables, as in the phrase "COT-tle-ston PIE." The 2nd line *must* consist wholly of a proper name of the correct rhythm. This may be a real person (O-lin S. PET-tin-gill), one in literature (LIT-tle-red- RID-ing-hood), or a scientific name (SOR-ex cin-E-re-us). Titles, if legitimate, may be used (GEN-er-al WASH-ing-ton). The 2nd line of the 2nd stanza must be a single six-syllable word in double dactyl rhythm (HET-er-o-SEX-u-al).

Here are several amusing examples of double dactyls that have appeared in *The Auklet:*

> Zippety zappity
> *Struthio camelus*
> Taxied along at the
> Speed of a jet.
> Something was wrong with it
> Aerodynamically;
> Sixty-five miles and it's
> Taxiing yet.

Brewster Award Winner
George H. Bartholomew
Studying temperatures,
Found, so I hear,
Speaking entirely
Physiologically,
Nothing replaces a
Cold can of beer.

Males in the family
Phalaropodidae
Sit on the eggs till the
Young bust the shell.
Females, because of this
Irregularity
Join up in flocks and raise
All kinds of Hell.

There is a species called
Puffinus puffinus.
"Surely a Puffin, of
Course," you will say.
No, it is found in the
Procellariidae.
That's why taxonomy's
All but passe´.

Opportunistical
Father Jim Mulligan
Taped all his sermons as
Well as his birds.
To his dismay, his new
Audiospectrograph
Showed all the music, but
None of the words.

Highly ingenious
Nicholas Collias
Wove out of grasses a
Weaverbird nest.

All the *Ploceus* were
Superdisconsolate
When they found out that his
Product was best.

DuPont executive
Crawford H. Greenewalt
Photographs hummingbirds
Crisply and clear;
Luck such as his, I find
Incomprehensible— ˙
All that my Brownie can
Get is a smear.

Kenneth C. Parkes, Editor
Richard C. Banks and Robert W. Storer
The Antic Alcid:
An Anthology of the Auklet
The American Ornithologists' Union, 1983

Corrosion Characteristics of Inconel

Frank LaQue is one of the great names in the field of corrosion of metals, having been associated with International Nickel Company (INCO) for many years. Among his many consulting activities in his retirement is one in connection with the Ocean Thermal Energy Conversion program. At a recent meeting of the OTEC Working Groups on Biofouling and Corrosion on the one hand and Heat Exchangers on the other, Frank told the following story from his earlier years:

It seems that a company that had been manufacturing aircraft engine exhaust manifolds during World War II found itself burdened with a substantial quantity of Inconel sheet at the cessation of hostilities. The bottom had suddenly dropped out of the market for high-performance aircraft engines, and the company was casting about for ways in which it could use up its inventory by making products in more immediate demand. One thought that crossed their mind was that they would use the Inconel sheet to make coffins. They wrote to INCO concerning the corrosion characteristics of Inconel in contact with soil. The letter was duly referred to Frank for his expert opinion. Soberly, he reported that the available data indicated that Inconel should be quite satisfactory for that service. But, as anyone who knows Frank might have guessed, he could not let it go at that. In a postscript, he suggested that in their advertizing campaign, they might draw attention to the heat-resisting qualities of Inconel and even use as a slogan, "Go to Hell in Inconel!"

The aircraft engine company predictably responded to this letter by writing INCO a testy letter to the effect that they didn't really think it was all that amusing. The official response of INCO to this seems not to be permanently recorded, but it should be noted that eventually Frank was made a vice president of INCO.

Kenneth J. Bell
Heat Transfer Engineering
Vol. 2 (3-4), p. 121, 1981

GILMANIA;

BEING A

Thesaurus

of Verbal and Pictorial Efforts, on the Part of divers
ORGANIC - CHYMISTS,

AT A

Brief Commentary,

upon their Experience in the

LABORATORIES

OF

The Iowa State College

OF AGRICULTURE AND MECHANIC ARTS.

Ames

M C M L I V

THE MAN OF CHEMISTRY: A Hitherto
Unpublished Section of the *Prologue* to GEOFFREY
CHAUCER'S *Canterbury Tales.*

Ther was al-so a man of chymistre,
I wot that HENRY GILMAN highte he;
He had grete lore of bodyes organeke,
And of the same ful longe colde he speke.
Of metall-carbon bondes and their wayes
He mighte discours for fourty nightes and dayes,
Ere of his lerneynge cam he to the ende;
And noon his sayinges rightly colde amende.
To him ech yeere a dele of clerkes ther came,
To lere of him the carbon-chymists game;
Ech clerk abood with him ten yeeres or so,
Then went his way with-outen wordes mo.
I mene, that this was trewe, but for the faster;
For some took fiftene yeeres to get a *Master.*

Auctor he was, and Editor as wel;
His *Treatis* did lyke very hoot-cakes sel,
And specialy at *Ames*, in *Iouay*,
This boke founde market large, it is no nay.

A mery lyf this Henry Gilman lad;
He alweys was in finest clooth y-clad;
His coot was butoned with butones thre,
Nat even butones he lat idle be.
Advances grete in chymistrye he maked.
Bifor his anger al his clerkes quaked.

THE CARBON CHEMIST; or, *Headaches Due to Organic Disturbances;* Being some ACCOMPT of the grievous ILLS Suffered by a certain STUDENT; Sung to the Vulgar Air, *"Strip Polka."*

OH, I was a carbon chemist when I came to I. S. C.,

And I swore I'd be a Doctor of Philosophy;

And I had so little foresight that I voiced the idle boast,

That I'd make it in nine quarters at the most.

I. S. C., I. S. C., I was thrown into your maw;

Ph. D., Ph. D., that I never, never saw;

Oh, I was a carbon chemist, and my hopes were running high;

But they fell, and here's the reason why.

Now they put me under Gilman, my research he
   would direct,
   This famed colleague of Grignard, held in wide
   respect;
On furan and free radicals his work is widely read,
   And he told me to start out on ethyllead.

Ethyllead, ethyllead, you made life a hell for me;
   Better dead, better dead I would definitely be;
You are toxic to the body and corrosive to the soul,
   And you drove me to drink pure ethanol.

AFTER wasting seven quarters on this compound weird
and quaint,
  Unto Uncle Henry Gilman I made loud complaint;
And he said, while meditatively he stroked his famous
chin,
  "Well, suppose you do some work on phenyltin."

Phenyltin, phenyltin, it was all a big mistake;
  What a sin—you can't win—every day three moles
  I'd make;
Of some important use for it I daily tried to think,
  But no soap, so I poured it down the sink.

On this God-forsaken compound I travailed eight
quarters more,
But everything I did had been done long before;
And when I complained to Gilman, Uncle Henry
said, "I think
That we'll let you do some work on butylzinc."

Butylzinc, butylzinc, it was profitless, I wis;
How I shrink from the brink of the butylzinc abyss!
It made me wish that I were back at work on phenyltin,
For I hate $(C_4H_9)_2Zn$.

WHAT I did with tolylsilver you are better left untold,
    As well as my affair with *iso*-amylgold;
The details of xylylyttrium I shall not here relate;
    They would put you in a melancholy state.

Metal preps, metal preps, I have done them, every one;
    Forty steps, forty steps in each metal prep I've run;
But the climax came when Uncle Henry said to me
    one day,
    "You can start on bornylneon right away."

WELL, my patience now is ended; I have taken all I
   can;
I am working in Nevada as a garbage man;
And I get two hundred monthly and will soon be
   getting three,
   Which is more than I got at I. S. C.

I. S. C., I. S. C., how you threw me for a loss;
   Henry G., Henry G., twenty years you were my
   boss;
Oh, I was a carbon chemist, now a garbage man am I,
   And now you know the reason why.

Bureaucracy

"Special Forms," the Lady Tells Him

By the Input/Output Window
Of an IBM 360
Sits an old man—bearded warrior.
Silent. Waiting for his program.

As a young man he submitted
90 cards—a *FORTRAN* program.
Walked away believing he would
In two hours receive his output.

Through the snow and wind returning.
Through the door; up to the window.
Calmly, with a pleasant smile, he
Asks the lady for his output.

"Turnaround," she tells him sharply,
"Turnaround is now six hours."
With her hand she indicates the
Blackboard on the wall behind her.

With a shrug he quits the window.
Turns around and climbs the stairwell.
Sits and studies for six hours.
Then returns—presents his check stub.

"Special forms," the lady tells him,
"Special forms have now been mounted

Even planets in their orbits
Halt when special forms are running.

"But," he questions, "I was promised
I'd receive my printout, if I'd
Wait six hours for its running."
"Special forms," the lady tells him.

Once again he shrugs his shoulders.
Buys himself a Dr. Pepper.
Buys a couple of frozen Twinkees.
Waits until the moon has risen.

Education comes not only
In the classroom from a teacher.
Mephistopheles grins as they
Tell him that his deck is missing.

Now an odd expression slowly
Spreads across his formerly calm face.
He sits down next to the window.
Sets his jaw. His eyes become glazed.

By the Input/Output Window
Of an IBM 360
Sits an old man—bearded warrior.
Silent. Waiting for his program.

Dan Nessett
Original contribution
to this volume

President Elwood Wheeler's Retrenchment Plan

"We're down to faculty, Elwood. There's nothing else left to cut."

Alan Fortran, Financial Vice President of Urban State University spoke to President Wheeler across a table littered with computer printout sheets.

"Things before people, Alan. We're out of things. The cupboard's bare. If we don't have a mild winter, we might have to ask everybody to bring in buckets of fuel oil just to keep the frost off."

"I hate to see people go."

"Look. With the state appropriation cut and the student enrollment not only down but shifting to new areas, we've got too many faculty members locked in the wrong places. Think of this place as a lifeboat, Elwood. We can't stay afloat without jettisoning some people."

"Well, when you put it that way, I can think of a few people whose departure wouldn't exactly be a great catastrophe."

"Who, for instance?"

"Henry Gizzard."

"Try again. Gizzard's tenured."

"*I* didn't give him tenure."

"Your predecessor has tied your hands on a lot of things, but don't mess with the tenure issue... Tenure protects faculty members so they can be free to make discoveries and announce controversial findings to the world without fear of getting canned."

"Gizzard? The only thing he's discovered in fifteen years was a parking ticket the security people put on his windshield, and he denounced me in the student newspaper as a bureaucratic jackass."

"You couldn't get him then, and you can't now."

"But this is a financial emergency."

"The State Board has handed down guidelines to handle just this sort of situation. Basically, it's last hired, first fired."

"That's starting at the cheap end."

As he spoke, a jet taking off from the municipal airport made a climbing turn over the campus. The roar of the engines rattled the windows and drowned out their conversation. He rubbed his chin slowly, then turned to Fortran and spoke deliberately.

"I think there's a way, Alan."

*   *   *

For the annual Christmas party, the dining hall was decked with twisted crepe paper streamers, alternately red and green, under which faculty members mingled and murmured while clutching styrofoam punch cups. Wheeler tapped the microphone several times and the resulting thuds caused everyone to turn toward him.

"Ladies and Gentlemen. May I have your attention, please? This is always a happy time when the magnificent holiday spirit can touch our lives and raise our spirits. On this occasion we traditionally show our appreciation for our fellow toilers in the vineyards by presenting Exemplary Faculty certificates. This year we have something extra. The Exemplary Faculty members will not only get the certificates, but, during the January semester break, when the rest of us are up here freezing our ears off, they will be down on the white sands of the Bahamas, courtesy of the University, all expenses paid!"

A gasp went up from the gathering.

"Yes, budget crisis be damned. Our priorities are people. Now, let's see who the lucky winners are. Vice President Fortran, may we start with the envelopes, please?"

Fortran handed him the first of a pile of manila interoffice envelopes. Wheeler tore it open with the fingernail of his pinky.

"The first winner is . . . . Henry Gizzard!"

"And now," said Wheeler, "may I have the next envelope, please?"

*   *   *

At noon on the last day of exams in January, President Wheeler stood at the departure gate in the low chain-link fence alongside of University Hangar, at the far end of Municipal Airport. A large twin-prop transport was parked on the concrete apron in front. From behind him, Wheeler heard a car door slam. He turned and saw Professor Gizzard get out of a taxi and come towards him. Gizzard was wearing a long, gray overcoat and a baseball cap. In one hand he carried a bulging pullman case, and over his other shoulder hung a set of golf clubs. Gizzard looked at the transport plane quizzically.

"What plane are we going in?"

"The big one."

"The C-46? Jeez, why don't we take a newer one?"

"We couldn't get everybody into a Beechcraft. There's twenty-seven in all, not counting the pilot."

"But I've never seen that old crate fly. Last time I took a trip out of here, it was all over the hangar floor in pieces."

"The Department of Aviation Technology has put more tender loving care into the plane than any bird you'll see around here."

*   *   *

Inside the flight operations shack next to the hangar, Professor Scott Andrews, veteran pilot and head of the Department of Aviation Technology, was leaning over the flight planning table when Wheeler walked in.

"How's the plane, Scott?" asked Wheeler.

"We did a performance check on both engines, sir, and we'll get take-off power with no sweat. The controls check out OK. Same with the auto pilot."

"What about the spot where you'll make your descent?"

Andrews reached into the shoulder pocket of his gray-green winter flying suit, took out a pencil, and drew a line along a large plastic protractor on the flight map.

"There it is, sir," said Andrews. "I've done some sky diving down there last summer. From there the plane has a straight shot out...."

He moved the pencil along the protractor out into the blue area, indicating the Atlantic Ocean.

"You know, Scott," said Wheeler, "some people would say that what is about to happen is an extreme measure. But a president in this kind of academic world has so few options he can control."

"Well, 'For everything, there is a time.' That's poetry... or is it the Bible?"

Wheeler watched as Andrews turned and crossed to the bench. On it rested two parachutes. Andrews bent down, hoisted the primary chute, and slipped it over his shoulder.

"Can I give you an hand on that?" asked Wheeler.

"No, don't!" shot back Andrews, then he caught himself. "I'm sorry, sir. I'm just very careful about my parachutes. The first rule of survival in this business is to rely entirely on yourself. That's lesson number one for the students in Introductory Sky Diving."

"You know, I could do without the heat from the boys in the legislature for giving college credit for jumping out of airplanes."

"That course is only an elective to give the kids a feel for the sky. The real business of aviation technology is real down-to-earth and practical. At least our graduates are getting jobs, and that's more than I can say for some departments I could mention."

"A good liberal arts education prepares a person for anything."

"It prepares them to appreciate our program, that's for sure. I tell you Mr. President, if we had more up-to-date equipment, our boys. . . and *girls* too!. . . would be ahead of everybody. We've gotten just about eveything we can out of that old C-46, anyway!. But when the budget picture brightens, you know where to put your chips."

"I've already assured you, if the plan works, your budget is safe next year."

"I've been thinking about that," said Andrews, pausing. "We'll need 10 percent more."

Wheeler looked hard at Andrews.

"I read you loud and clear."

*   *   *

When Wheeler arrived back at the departure gate, the rest of the group had gathered and was engaged in gay banter. Miss Templetoe spotted him and came over. At that moment Andrews hailed the group as he came up to the departure gate.

"All right, everybody! On Board! Grab your stuff and lug it with you. This isn't Eastern Airlines!"

One by one the group filed after Andrews through the departure gate and up the ramp into the cavernous interior of the plane. When they were all on board, Andrews closed the door. Two students from the Department of Aviation Technology wheeled away the ramp.

First one engine, then the other coughed and caught with a deafening roar, blasting back clouds of blue smoke. From the cockpit window, Andrews looked down at Wheeler and gave him the thumbs-up signal. Wheeler returned the gesture.

The plane slowly taxied down to the far end of the runway. Wheeler raised his hand to shield his eyes from the low winter sun. He stood there as if in salute as the plane revved up its engines, churned down the runway, and rose ponderously into the air.

*   *   *

From his expansive office window high atop the administration building, President Wheeler looked down at the evening rush-hour traffic. He was alone, waiting. He periodically looked over at the telephone on his desk. Finally, it rang.

"Hello, Wheeler, here," he said, grabbing the receiver off the cradle.

There was a confused jumble of buzzes and electronic tones, plus faint bits of feminine talk in a strange tongue. In a moment, a raspy, familiar voice came through.

"Is that you, Prexy?" Wheeler stared blankly ahead.

"Who is this?" he said.

"This is Henry."

"*Henry*?"

"Henry Gizzard, for Christ's sake."

"Where are you calling from?"

"We're in . . . what's the name of this place?. . . Just a second... Alice Templetoe is getting on another extension."

"Hello, President Wheeler," came the lilting sound of Miss Templetoe's voice. "We're in Balaquendos, the capital of a new island republic in the Caribbean. But," her voice dropped, "we have startling news for you."

"This has been startling enough already," said Wheeler.

"You see, Andrews is not with us."

"Oh!"

"No, we have no idea where he is!"

"What happened?" asked Wheeler.

"Well," Miss Templetoe said, "it was as we were eating our delicious lunch and were having a real party time of it, when we noticed that he made his way down the aisle to the back of the plane."

"I thought he was going to the can," interjected Gizzard. "Pretty soon there was this sound of rushing air, and when we went back to check, the back door was open. He wasn't in the can, or anyplace."

"You don't suppose," said Miss Templetoe, "that in the festive spirit, he opened the wrong door?"

"What did you do then?" asked Wheeler.

"It was all very terrifying, and it took us a long time to do anything," Miss Templetoe said.

"We finally broke into the front cabin," said Gizzard, "and found the controls on autopilot heading us straight out to sea."

"And then?"

"Well, you know Norton in Philosophy?" continued Gizzard, "He had been in Liberators in the European theatre during World War II."

"Thirty-five missions!" Miss Templetoe darted in.

"And Wurtz in History had flown Gooney Birds for the Air Rescue Service. They're by the same maker as that one. He said, 'It's like riding a bicycle—once you know how, you never forget.'"

"And Jennings in Physics," said Miss Templetoe. "He had been in fighters so of course we couldn't let him near the controls. But he figured out the radio very nicely."

"Let me get this straight," said Wheeler. "You set up a committee to fly the plane?"

"Something like that."

"I never heard of a faculty committee accomplishing anything."

"The task was well-defined," said Gizzard.

"I felt like Helen Hayes in those Airport movies," said Miss Templetoe.

"Words cannot express my feelings at this moment," said Wheeler after a long pause. "Tell me, what are you up to now?"

"Most of us are in the hotel, the big one in the center of town," said Gizzard. "We hear there is a beautiful beach not far from here, and a couple of golf courses."

"We are going to look them over in the morning," said Miss Templetoe. "Right now, Gomez of Romance Languages is negotiating with the locals about the plane."

"What about the plane?"

"It continued rolling off the end of the runway and one wing touched the ground. Fortunately, it was all mud and the plane just settled in nicely. Gomez thinks we can get enough in salvage to pay our hotel and first-class tickets home at the end of the week."

*   *   *

"I cannot say too sincerely that I am overjoyed that you are all safe and happy," said Wheeler.

"We'll be seeing you, then, and thank you for the wonderful idea for the trip to this enchanting place," said Miss Templetoe.

"Don't mention it," said Wheeler.

He put the receiver down and stared ahead for a long time. Finally he picked up a pad of paper and began writing. The telephone rang again. Wheeler picked it up.

"Wheeler here."

"The Eagle has roosted," said a muffled voice.

"Cut the code, Andrews, and tell me where the hell are you?"

"I'm in the general area I picked out. But I missed the timing of the jump by a little bit and the wind took me over into the woods. I was hung up in a tree until the volunteer fire department came and got me down."

"I'm glad at least that the volunteer fire department is on the ball."

"Listen, Mr. President. It's now six-thirty. There was fuel on board only for five hours. So I guess it's all over by now."

"Yes, it's all over now," responded Wheeler, dryly.

"Say, there's a reporter hanging around from the local newspaper, and I want to make sure our stories are absolutely together, right?"

"As a matter of fact, I was just preparing a press release of my own."

"Oh? What does it say?"

"It says, 'Due to extreme financial exigency at the Urban State University, President Elwood Wheeler announced today that the Department of Aviation Technology will be abolished at the end of the current academic year.'"

"Mr. President," stammered Andrews, "what ever do you mean?"

"It means that we are selling the planes, and sweeping out the nuts and bolts—you included."

"But we were just making such great headway through the bastions of the Liberal Arts!" said Andrews.

"Look, I have had a convincing demonstration that a thorough grounding in the Liberal Arts prepares a person to handle anything," said Wheeler. "Andrews, you wouldn't believe what those bastions have done!"

John B. Haney
AAUP Bulletin
p. 17, February 1977

Ballade of the Reluctant Deans

Why does a man become a dean?
    Long have I pondered the question too.
Three are the reasons I have seen.
    Names they are called would support this view:
Sinister fate, as the villain true,
    Destines some to a doom foreseen;
Vainly their heritage they rue,
    Those who are born for the role of dean.
Others strive from the ranks to rise;
    Far beyond deanship they aspire:
The presidency their cherished prize.
    Power and glory they desire;

Deans though they be, they would rise still higher,
    Holding as naught achievement mean.
Bitter and frustrate, they retire,
    Viewing with scorn their role of dean.

Few are the deans on whom are thrust
    Honors they merit but do not seek.
Like wise men whom Plato describes as just,
    They consent to rule, lest the deans be weak.
The committeed life they endure, and speak
    Wisely and well, with pontifical mien.
Paragons they, though not unique,
    Who nobly adorn the role of dean.

Envoi

Regent princes, I pray you, heed
    Faculty voices and judgment keen;
Contravene fate and frustrate greed:
    Choose the "third man" for role of dean.

Elizabeth M. Kerr
Bulletin of the AAUP
Vol. 42, 773, 1956

The Tin Men

In his 1965 jest on computer programming, Michael Frayn examined the William Morris Institute of Automation Research where Hugh Rowe was attempting a computer-written novel and Goldwasser's Newspaaper Department was collecting clippings to program standard stories such as "Paralyzed Girl Determined to Dance Again" . . .

In the Ethics Department, Macintosh had concentrated all the department's efforts on the Samaritan program. The simplest and purest form of the ethical situation, as he saw it, was the one in which two people were aboard a raft that would support only one of them, and he was trying to build a machine that would offer a coherent ethical behavior pattern under these circumstances. It was not easy. His first attempt, Samaritan I, had lacked in discrimination in the self sacrifice. He had now developed Samaritan II, which would sacrifice itself only for an organism at least as complicated as itself.

The raft stopped, revolving slowly, a few inches above the water.

"Drop it," cried Macintosh.

The raft hit the water with a sharp report. Sinson and Samaritan sat perfectly still. Gradually the raft settled in the water, until a thin tide began to wash over the top of it. At once Samaritan leaned forward and seized Sinson's head. In four neat movements it measured the size of his skull, then paused, computing. Then, with a decisive click, it rolled sideways off the raft and sank without hesitation to the bottom of the tank.

"Save it, Lord," boomed Macintosh to a young man waiting on the side of the tank in a swimmming costume. Lord dived in and attached a rope to the sunken Samaritan.

"Why don't you tie the rope to it before it goes overboard?" asked Goldwasser.

"I don't want it to know that it's going to be saved. It would invalidate its decision to sacrifice itself."

"But how would it know?"

"Oh, these Samaritan IIs are canny little beggars. Sometimes I think they understand every damned word you say to them."

"They're far too simple, Macintosh . . ."

"No, no. They come to trust you. So every now and then I leave one of them in instead of fishing it out. To show the others I mean business. I've written off two this week."

He leaned over the rail and shouted: "Oh, Lord! Is it all right? Load it up for the next run, then."

"What are you doing now?" said Goldwasser.

"We're starting a new series to test its behavior with simpler organisms."

Samaritan II came back up to the gantry, winched by the crane. There was something about its visible dials and displays that struck Goldwasser.

"Doesn't it look a bit sanctimonious to you?" he asked Macintosh.

"Aye, it always does after it's gone over the side. It's a minor defect. We'll get it right in development."

"But Macintosh, if it enjoys sacrificing itself it's not taking an ethical decision at all, is it?"

"I don't see why it shouldn't enjoy doing right."

"But if it's enjoyable it's not self-sacrifice."

"By God, Goldwasser, you're a real puritan! If a thing's right it's right, and if you enjoy doing it so much the better."

"It may be right. But for God's sake, Macintosh, it's not ethically interesting!

*   *   *

A large sandbag was winched up to the gantry, and placed side by side with the Samaritan on the raft.

"Launch her!" shouted Macintosh.

The raft swung out over the water, and was lowered steadily away. The resonant silence fell once more.

"There's a logical flaw here," said Goldwasser suddenly, and his voice boomed about the roof.

"Drop her!" shouted Macintosh.

Samaritan and the sandbag looked at one another impassively as the raft settled. When the deck was awash Samaritan seized the sandbag and attempted to measure the size of its skull. It attempted its four neat movements, frustrated by the un-skull-like shape of the sack, then paused, drew its conclusions, uttered a thoughtful whirr, and became completely still.

"Good lad," said Macintosh under his breath.

The deck of the raft was now completely submerged. Gradually the water rose around Samaritan and the sandbag as they sat stoically accepting their fate. The sandbag was the first to disappear. Then, with a last look of silent martyrdom, Samaritan vanished too. The bulging, shrinking, dark, refracted shape beneath the water went steadily down to the bottom.

"Well, I hope that meets all your objections to Samaritan I," said Macintosh. "You see it didn't even attempt to sacrifice itself for the sandbag."

"I see that," said Goldwasser. "But Macintosh, the only result was that they both went to the bottom."

"Oh, Goldwasser," said Macintosh, "you're just a rotten cynic."

If only I could get away from my damned Samaritans for a week or two! Did I ever tell you about my idea for programming a computer to write pornographic novels? Well, I sometimes wonder if you couldn't program machines to perform a great deal of human sexual behavior. It would save a lot of labor."

"Yes," said Rowe. "Yes."

"At any rate in the early stages. On the same principle you might also program machines to go through the initial conversational moves two people make when they first meet. They're always standardized, like chess openings. You could select your gambit, then go away and make the tea while the machine played it, and come back and pick up the conversation when it started to become interesting."

"Yes," said Rowe.

"Its breaks my heart to see that wing going empty when there's so much to be done. I mean, let's accept—and I owe this suggestion to my good friend Goldwasser—that all ethical systems are ossified, in which case all operations within an ethical system can be performed by computer. I should be designing circuits to demonstrate what happens when one ossified system, say a Christian one, comes into contact with another ossified system, say a liberal agnostic one. And what happens when two computers with incompatible systems try to program a third between them.

"Ah, Rowe, Rowe, Rowe! Doesn't it move you to contemplate the great areas of life that have ossified, where activity has been reduced to the manipulation of a finite range of variables? The pity and the terror of it, Rowe! These vast petrified forests are our rightful domain. They are waiting helplessly to be brought under the efficient, benevolent rule of the kindly computer.

"Take the field of religious devotions. What computer man can survey devotional practice without lifting up his heart and thanking God for sending him such a prize? When we are called in to write a program for automated devotion—as we laid down by the Thirty-Nine Articles or the Holy Office without fear of heterodoxy. But in fifteen or twenty years' time we shall be writing programs for praying. The subjects and sentiments tend to come in a fairly limited range."

"Ah," said Rowe, "there's a difference between a man and a machine when it comes to praying."

"Aye. The machine would do it better. It wouldn't pray for things it oughtn't to pray for, and its thoughts wouldn't wander."

"Y-e-e-s. But the computer saying the words wouldn't be the same...."

"Oh, I don't know. If the words 'O Lord, bless the Queen and her Ministers' are going to produce any tangible effects on the Government, it can't matter who or what says them, can it?"

"Y-e-e-s, I see that. But if people say the words, they *mean* them."

"So does the computer. Or at any rate, it would take a damned complicated computer to say the words *without* meaning them. I mean, what do we mean by 'mean'? If we want to know whether a person or a computer *means* 'O Lord, bless the Queen and her Ministers,' we look to see whether it's grinning insincerely or ironically as it says the words. We try to find out whether it belongs to the Communist Party. We observe whether it simultaneously passes notes about lunch or fornication. If it passes all the tests of this sort, what other tests are there for telling if it means what it says? All the computers in my department, at any rate, would pray with great sincerity and single-mindedness. They're devout wee things, computers."

"Y-e-e-s. But I take it you don't believe in a God who hears and answers prayers?"

"That's not my end of the business. I'm just concerned with getting the praying done with maximum efficiency and minimum labor output."

"But, Macintosh, if you're as cynical as that, what do you think the difference between the human and the computer is?"

"I'm not absolutely certain, Rowe. I'm inclined to rule out what you might call 'soul.' I should think that in time we could teach computer to be overwhelmed by the sound of Bach or the sight of another computer, to distinguish between good sonnets and bad, and to utter uplifting sentiments at the sight of the Matterhorn or the sunset. Obviously it's not the faculty of choice; computer can be programmed to choose, just as humans do, rationally, anti-rationally, at random, or by any combination of the three.

"Now you and I, Rowe, as practical men, would say that the only useful working distinction between people and computers is that computers can choose only from among a finite range of variables, whereas people can specify for themselves the range of variables from which they will choose. But it's a wee bit tenuous, because it depends simply on the complexity of the human neuromechanism. I suppose one day we shall build a computer as complex, and it will begin to specify its own limitations."

"Of course," said Rowe.

"I dare say, in the long run, the distinction will prove to be an economic one, you know. In the same way that it's cheaper to use a computer for finite intellectual tasks, there may well come a point beyond which the open-endedness of a once-only job would require a computer so complex and so specialized that it would be cheaper to use a human being. Aye, I believe there will always remain certain areas on the fringes of the petrified forest where original thoughts have to be thought, and original juxapositions of ideas made, and new meanings and possibilities seen. I shouldn't be surprised if it proved more economical to use people rather than human-like computers to work these areas. As a computer specialist I naturally regret it. But as a human being I must admit I get a certain sneaking pleasure from believing that there are jobs to be done that are worthy of the human mind. The awe and terror of it, Rowe! The pity and the grandeur! Do you see what I mean?"

"Yes," said Rowe, as they got up from their table in the Tea Room to return to their laboratories. "Yes. Yes."

He felt slightly exhausted, but also exhilarated. What a stupendous countryside of propositions they had marched through! Petrified forests! Infinite ranges of variables! Computers really deeply religious things! Sincerity! Choice! Complexity! Human beings all right after all! He felt a great sense of comradeship with Macintosh, who had borne him company on this extraordinary journey. An extract from a review hung in lights in the sky above the infinite forests and the petrified ranges, and he felt both generously pleased—and moved that he was generously pleased—to see that for once the celestial reviewer referred not to him but to Macintosh. "Macintosh..." it said "...A wonderful listener...brings out all the dazzling intellectual curiosity in Rowe..."

Michael Frayn
The Tin Men
Collins, London, 1965

Thermodynamics and Administration

The discovery by Barby that the same laws governing physical chemistry also apply to politics have inspired the author to investigate the laws of thermodynamics applied to University Administration, in general, and to committees, in particular.

A committee may be regarded as a closed system consisting of N people. The total amount of useful work (ΔW) done by a committee is given by the summation of the useful work done by individual members:

$$\Delta W = n_1 w_1 + n_2 + ... n_N w_N \tag{1}$$

$$= \sum_0^N nw = N \sum_0^N w \tag{2}$$

Now the amount of useful work is related to the change in free energy by:

$$-\Delta G = \Delta W - P \Delta V \tag{3}$$

where P is the pressure, in this case exerted by the Administration and ΔV is the change in volume of a committee during a meeting (reaction). The administrative pressure exerted on a committee has no net effect upon its equilibrium.[2] This means that there is no volume change as a result of a meeting (reaction); i.e.:

$$\Delta V = 0 \qquad\qquad (4)$$

hence:

$$P\Delta V = 0 \qquad\qquad (5)$$

Substituting Equation (5) into Equation (3) gives Equation (6):

$$-\Delta G = \Delta W \qquad\qquad (6)$$

Since a committee takes place in the gas phase:

$$\Delta G = -RT \ln K_P \qquad\qquad (7)$$

where R is the gas constant, T is the temperature of the committee (usually arbitrarily designated 'hot' or 'cold') and K_P is the equilibrium constant in terms of pressure. Now $K_P = 1$, therefore from Equation (7), $\Delta G = 0$, i.e., a committee is at equilibrium. Hence from Equation (6), $\Delta W = 0$. This means that no useful work is done by a committee. This is borne out in practice—committees are notorious for spending hours making Minutes. Since $\Delta W = 0$, it can be seen from Equations (1) and (2) that any useful positive work done by individual members in a committee is cancelled out by negative useful (i.e., useless) work done by other members.

Further information about a committee may be gained by considering the first law of thermodynamics, which states that in any closed system energy is conserved. This may be represented mathematically by:

$$\Delta E = Q - \Delta W \qquad\qquad (8)$$

and:

$$\Delta E = E_2 - E_1$$

where ΔE is the change in internal energy of a committee; E_1 and E_2 refer to the energy at the beginning and end of the meeting ($E_2 > E_1$); Q is the heat absorbed by a committee, usually in the form of hot air emitted by the members; and ΔW

is the amount of useful work done by a committee. In this situation $\Delta W = 0$, hence:

$$\Delta E = Q \tag{9}$$

It is well known that for any system:

$$\Delta H = \Delta E + RT\Delta n \tag{10}$$

where ΔH is change of enthalpy, R is the gas constant, T is the temperature of the meeting and Δn is the change in the number of committee members during a meeting. Since $\Delta n = 0$:

$$\Delta H = \Delta E \tag{11}$$

$$= Q \quad \text{from Equation (9)} \tag{12}$$

But the change in free energy is related to the change in enthalpy by the well-known Gibbs' Equation (13):

$$\Delta G = \Delta H - T\Delta S \tag{13}$$

where ΔS is the change in entropy of the committee. Now $\Delta G = 0$, hence:

$$\Delta H = T\Delta S \tag{14}$$

i.e.:

$$Q = T \Delta S \tag{15}$$

Since both Q and T are positive, ΔS is positive and so Clausius' statement of the second law of thermodynamics is again verified: "In any isolated thermodynamic system, entropy tends to increase to a maximum." Entropy is that magic term concerned with the order/disorder of a system. In simple terms, Equation (15) means that as a committee meeting proceeds, disorder occurs.

Now the rate (k) at which a committee meeting proceeds is given by the Arrhenius equation:

$$k = A \exp (-E/RT) \tag{16}$$

where E is the activation energy, i.e., the energy required to get the committee to meet, and A is the Arrhenius Administrative factor. Equation (16) can be written as:

$$\log k = \log A - E/2.3RT \qquad (17)$$

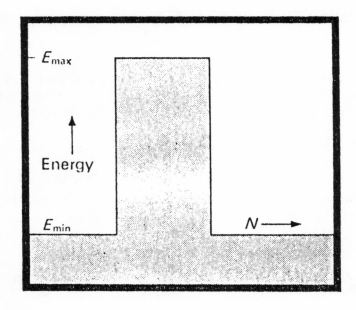

Fig. 1. Energy distribution of committee.

It is a general personal observation that as the temperature of the meeting is lowered, then the meeting proceeds faster. This means that the activation energy is negative—no energy is required to get a committee to meet; individual members love to meet and form a committee at every available opportunity. The Administrative factor also plays an important role in committee procedure; it can be shown from Equation (17) that the more administrators on a committee (i.e., increase in log A), the faster a committee will proceed.

A final word about energy distribution. Figure 1 shows a typical energy profile of any committee. Most of the members reside on the committee with the minimum amount of energy

(E_{min}). The number of members using the maximum amount of energy (E_{max}) are few: the chairman, secretary, and the administrator.

Conclusions

- A committee will do no useful work
- As a committee meeting proceeds, disorder occurs
- A committee has a negative activation energy
- The more administrators on a committee, the faster a committee will proceed
- The majority of the members of a committee reside there with the minimum amount of energy

References

1. D. Barby, *New Scient.*, 1970, **64,** 70.
2. Personal observation.

A. J. Monty White
Chemistry in Britain
Vol. 13(4), 150, 1977

The Chair

When the Department had finished its meeting,
Half of us went to have lunch at an Inn,
And while we were busily eating and drinking,
Somebody said, and he stuck out his chin:
"Do you wonder arrangements are always behind
When the silly old whatsit can't make up his mind?"

"We haven't much longer to stick the old fool!"
In only two sessions, he reckonded with joy,
The Faculty, Senate, Department and School
Would hold sherry parties to praise the Old Boy.
And when they'd all wittered about the Professor,
Whom did we think they would choose as successor?

We needed a person with some sort of name,
To put the Department back on to the map,
Who's known in his subject, with nation-wide fame.
We suddenly thought that we knew just the chap -
The man of the moment, Sir James Absentee,
This year's FRS* with this week's KCB[†].

There's a man who would bring us the power of decision—
Except when he's visiting Rome or New Delhi
Or taking the chair on a Royal Commission,
Advising the State, or appearing on Telly.
Not Someone not there simply isn't a starter;
We'd be far better off with young Rex Imperator.

He won't leave you feeling abandoned to fate,
But tell you precisely just what are your tasks.
He'll come very early and work on till late,
Seeing you do things the way that he asks.
With him you will not be left out in the cold
If you stand to attention and do as you're told.

The Dean might be sure there's an excellent boss
Among the young hopefuls who've not won their spurs,
But the rest would be scared he could be a dead loss,
And then we'd be saddled for twenty-odd years,
So, Stopgap from You-Know might get their attention;
He has a few years before drawing his pensions!

Then we tried thinking who might have been missed.
There were some very stodgy and some rather weird.
It made, in the end, a formidable list,
And the more we considered, the less we were cheered.
It's really a shame the Old Man has to go!
There's a lot to be said for the Devil you know!

Douglas E. Kidder
From *Verses Bright and Beautiful*
University of Bristol, 1983

*Fellow of the Royal Society.
[†]Knight Commander (of the Order) of the Bath.

Minutes of the Tissue Committee

The Tissue Committee of the Bucolic Springs General Hospital held its regular monthly meeting on March 26 at 12:30 PM. The chairman, Dr. N. E. Utral (Omphalology) presided. Others present were Drs. E. X. Cise (Surgery), P. Elvik (Gynecology), P. Lain (General Practice), and S. R. Donnik (Pathology). A selected verbatim report follows:

Utral: Let's get started. The first case is that of a 28-year-old white woman who had a hysterectomy for irregular bleeding.

Donnik: Pathologically, this was an essentially normal uterus.

Utral: There were no consultations. The case is open for discussion.

Elvik: Many times there are excellent clinical indications for hysterectomy when Donnik can't find anything. We shouldn't condemn a gynecologist just because the pathologist can't see gross or microscopic lesions.

Utral: The case was not done by a gynecologist, but by a general surgeon, Dr. S. Lash.

Elvik: That's different. In my view, general surgeons shouldn't do hysterectomies. They don't have the background to evaluate the total physiological picture of the female patient. Only well-qualified gynecologists should do hysterectomies.

Cise: I disagree. A general surgeon is well able to take out a uterus.

Elvik: I move this case be referred to the Executive Committee for consideration and possible disciplinary action.

Donnik: I second the motion.

Drs. Elvik and Donnik favored the motion. Drs. Utral, Cise, and Lain opposed it. Motion failed. Case filed with no further consideration indicated.

Utral: The second case is that of a 14-year-old girl with pain in the lower quadrant, irregular menses, and a 14,000 white cell count. An appendectomy was done.

Donnik: Pathologically, this was a normal appendix.

Cise: Well, nuts, that could happen to any good surgeon. It's just one of those things.

Utral: The operation was not performedby a general surgeon, but by a gynecologist, Dr. B. Urgeon.

Cise: That's different. In my view, gynecologists shouldn't do appendectomies. They don't have the background to evaluate the total surgical picture.

Elvik: I disagree. A well-trained gynecologist can handle a routine appendectomy easily. Anyway, on reviewing the chart it seems that Dr. Urgeon was considering ectopic pregnancy in his differential diagnosis.

Cise: I move this case be referred to the Executive Committee for consideration and possible disciplinary action.

Donnik: I second the motion.

Drs. Cise and Donnik favored the motion. Drs. Utral, Elvik, and Lain opposed it. Motion failed. Case filed with no further consideration indicated.

Utral: The last case is that of a 46-year-old white man whose infected, ingrown toenail was excised. We are considering this case because no consultation was obtained before surgery.

Donnik: Pathologically, this was a normal toenail.

Cise: That's a lot of baloney. Any good surgeon should be able to take off a toenail without consultation.

Utral: The operation was not performed by a surgeon, but by a general practitioner, Dr. F. Umble.

Cise: That's different. In my view, GPs should not attempt complicated surgery such as this. They

	don't have the background to evaluate and cope with the total surgical picture.
Lain:	Now, wait a minute, fellows, you've got to leave some minor surgery for general practitioners to do in this hospital. Umble is a good man on tonails.
Elvik:	I agree with Dr. Cise. Only specialists should do special surgery in this hospital. I move this case be referred to the Executive Committee for consideration and possible disciplinary action.
Donnik:	I second the motion.

Drs. Utral, Cise, and Donnik favored the motion. Dr. Lain opposed it. Case referred to the Executive Committee.

The Tissue Committee adjourned at 1 PM.

A. G. Foraker
JAMA
Vol. 187, p. 234, March 14, 1964

Fellowshipmanship

A Concise Manual on How To Be a Specialist Without Studying

We begin with the statement of the insignia of the College for Fellowshipmen: *Qui s'excuse, s'accuse.*

The Approach to General Practitioners

It is well at the outset to choose the image one intends to present to general practitioners, with the intention of putting the GP at a psychological disadvantage. Whatever role is adopted, the following rules must always apply:

1. The Fellowshipman must always appear to be fond of the general practitioner, in the same way as good citizens are fond of their dogs, or the Mounties of their horses.

2. The Fellowshipman must never openly disagree with the general practitioner on matters pertaining to medicine because there are far more effective ways of demonstrating error and because the general practitioner is sometimes right. (This latter possibility must never be admitted openly.)

Mannerization I

Some nondescript and totally meaningless mannerism should be carefully developed, such as a tut-tutting noise made with the tongue against the teeth, or a significant but ambiguous raising of the eyebrows. Such mannerisms are invaluable in circumstances where a categorical 'yes or 'no' might lead one to commit oneself to a definite opinion.

It is generally wise to ask the general practitioner to outline the history before going to the patient, so that it has to be recounted from memory without the use of notes. If the history is presented in a sketchy manner, the Fellowshipman should pursue it to great lengths. If, on the other hand, the history is presented efficently and clearly, the Fellowshipman should yawn, shift from foot to foot, examine his or her nails, and make it clear that it all seems a waste of time. A number of history-devitalizing questions can be used to interrupt and break up the flow of the general practitioner's presentation: "Just a minute. Was there a gag reflex present at that time?" or "I suppose the father wasn't a glass blower, was he?" or "Is the patient red–green color blind?"

The Presentation of the Diagnosis

This is the crucial point of the consultation. This is the pivot of the Fellowshipman's activity. Here, the whole approach should rise to a crescendo, the peak of which is the pronouncement of the diagnosis, and all that may follow should be diminuendo in an atmosphere of glory and adulation. The opening gambit in presenting the diagnosis should be calculated to intrigue and mystify. It can be of two kinds:

A. The Fellowshipman leans back, lights a cigaret or cigar, and then says, "Well, it isn't a Weir-Westergrand-Jones syndrome." Any questioning of such a diagnosis can be passed off since the Fellowshipman has just said that this is not the diagnosis in any case.

B. The second gambit is "Obviousmisting." The Fellowshipman begins by uttering some such remark as: "It is of course essential to note that this patient has normal vision." This gives the general practitioner something very puzzling to consider during the rest of the discussion.

Circumstances will dictate the final method of presenting the ultimate diagnosis. Under no circumstances should a positive diagnosis be given. In fact, the more obvious it is, the more it should be obscured. Thus, it is not permissible in the case of a patient who has consolidation of his right lower lobe, a fever of 104°F and pneumococci in his sputum to make a diagnosis of lobar pneumonia. The double negative diagnosis may be used to some effect. Thus: "It is impossible to say that his patient does not have pneumonia." The Fellowshipman may then select a special investigation (irrelevant as it may be), turn to the general practitioner and ask him for the results of this procedure—with the implication that if this had been done previously, there would have been no need to disturb the Fellowshipman during Thursday afternoon golf.

Mannerization II

A negative mannerism should also be cultivated. As an example, we may cite that developed by Garth Pindermoss, one of our honored graduates. Pindermoss would entice a fellow specialist into describing a recent operation he had performed. At a crucial point he would ask an apparently simple question, such as: "Did you undersew the superior inverted flap?" Whether the specialist replies yes or no, Pindermoss would then respond with the full negative mannerism. This consists of a half closing of the eyes, a partial frown, a slight shaking of the head, and the sound produced by drawing the breath sharply through pursed lips. This whole movement was beautifully executed by Pindermoss and could generally be guaranteed to have a paralyzing effect on the other specialist.

During the subsequent course of the description of the operation, it was never necessary for Pindermoss to re-execute the entire maneuver, but at any time he could engender a similar emotional catastrophe in his oponent by making use of any one of the separate parts, such as half-closing the eyes.

If the argument is progressing to the Fellowshipman's disadvantage, a number of moves are available. For example, the Fellowshipman first lulls the specialist into a sense of false security by seeming to be persuaded. This creates a favorable impression of impartiality and scientific open-mindedness. Then the following maneuver is executed:

Fellowshipman: Very interesting. A new approach! It
                 totally invalidates the fundamental work of
                 Rosenkran and Gildenstern.
Specialist: How do you mean?
Fellowshipman: Well, it's obvious, isn't it. (Pause.) You
                 don't mean to tell me that you are putting
                 forward this theory without comparing it to
                 their work?

At this point it is very simple to show that the specialist has never given adequate thought to the thesis. The faith of the audience in the specialist is completely shaken, not only in relation to the present argument, but in relation to the general standard of reasoning.

Ian Rose
Canadian Medical Association Journal
Vol. 87, pp. 1232–1235, Dec. 8, 1962

Tested Techniques for Avoiding Responsibility

Never be responsible for a project that fails! Practically, that means you must never be responsible for any project. Here's how to do it without being obvious.

Almost all individuals in our society seek that easily recognized but undefinable condition known as success. This article will attempt to provide techniques for success gathered from observations of individuals at all levels in organizations of all sizes.

The Principles are Simple

As with all great relationships in science, the principles are basically simple. The key fact is that in any organization there are many more projects that fail than succeed. Since it is difficult to know in advance just which projects will turn out which way, the secret of success lies in not being responsible for any of them. The royal road to success, therefore, is learning to avoid responsibility while not appearing to do so.

Levels of Competence

As with other forms of art, there are different levels of competence. The higher the level of responsibility, the greater the skill required to handle its avoidance properly. There are no sharp lines between levels of skill, but for convenience of discussion, they will be identified as follows:

BAAR   Beginner at Avoiding Responsibility
MAAR   Master at Avoiding Responsibility
PHAAR Pretty Hep at Avoiding Responsibility
AAAR   Artist at Avoiding Responsibility

These ratings are earned solely by how closely the individual approaches the ultimate objective, which is: getting complete credit for a project if it succeeds, while completely avoiding responsibility if it fails.

To qualify for a rating in responsibility avoidance (RA), the techniques used by the individual must not be obvious. The unskilled creature who uncouthly says, "I'm not going to stick my neck out!" is not eligible.

Organization Is Unnatural!

There is one enemy to be alert for. This is the Eager Beaver (EB), someone who is always trying to introduce procedures that practically force the making and implementing of decisions. As every technically trained worker knows, organization is contrary to the Law of Entropy, and therefore unnatural. It is often pathetic to see how hard people strive against this fundamental law of nature, but we must resist the temptation to be associated with anything so unnatural.

The Eager Beaver is easily identified and adequately handled by isolation. Actually, there is a serious problem only when the EB owns the company or manages a semi-independent unit. In other cases, even though the EB has somehow or other achieved a position of power in the organizational structure, the EB's tenure is usually temporary since eventually EBs comes up against advanced practitioners of RA and succumb from frustration.

Learning the Techniques

The techniques to be described are best learned in the order shown, as some forms of RA build on more primitive forms. It is useful to have a repertoire of techniques, not only to permit selection of the best one for a given situation—but shifting and combining them occasionally confuses the opposition.

The basic beginner's ploy is the petty excuse. For example, "I couldn't do the experiment, sir, because somebody took my beaker." Please note that it does not matter that there are hundreds of beakers in the stockroom. An obstacle has arisen, and that justifies abandonment of the assigned task. Of course this techniques has a certain lack of exclusivity since practically everyone is using it to some extent. Fundamentally, it consists of an individual's pursuing an assigned task only to the point of finding an obstacle big enough to justify failure to the superior. The failed one then points out how great was the effort, and only the cruelest superior would be critical of someone who tried so hard when confronting so insurmountable an obstacle.

Even a Small Obstacle Can Stop You

This technique has the advantage of illustrating the differences in the degree of competence of the individual. Everybody has a characteristically sized obstacle that stops one. The skill of the indivdual in RA can be measured by how small an obstacle can be used. An AAAR can justify stopping over pebbles, whereas some foolish Eager Beaver will find ways over, around, or through mountains to get the job done. Frightening fellows to have around sometimes; really rather like bloodhounds.

A simple RA technique that is almost invariably successful is that of the minor illness. "Let's call off that staff meeting this afternoon, fellows. I have a terrible headache and want to be sharp for those decisions." Or, "Better cancel my plane ticket to see Jones about that complaint. I feel a cold coming on. See if Bill can make it." Of course, if you are a person of few words, just don't show up for an unpleasant situation; have your wife (or husband) call in for you. This is somewhat more effective since it shows that you are so ill you can hardly reach the telephone.

The most difficult of the Beginner's techniques is that of developing one (or several) undesirable personality traits. Two that are especially recommended are quarrelsomeness and lack of cooperation; they are particularly effective when used together. Sticking to a position regardless of the facts is useful, and as well as having a reputation for never getting a job done on time. As with porcupines, others will leave you alone, and that includes assigning responsibility to you. Absence of friends is a small price to pay for such an achievement.

Now let us proceed to those techniques that earn for their users the rating of Master. These techniques all involve an addiional dimension over those used by Beginners. Their beauty lies in giving the impression that the practitioner is actually *taking* rather than avoiding responsibility! This group is distinguished from those practicing still more advanced techniques by their one shortcoming: the responsibility that the individual avoids tends to greatly inconvenience colleagues or subordinates.

The Business of Being Busy

The most common of the master-level techniques is that of being "too busy." This requires careful distinction from actually being "busy," which is an Eager Beaver trait. Being occasionally too busy does not qualify either—the situation must be chronic. In cases where there is some doubt, an infallible test can be applied. If a person is busy with the most important tasks one is qualified to handle, that person is an imposter. The true MAAR can be distinguished by being so busy with minor tasks that he or she can't get to his or her important ones. Also, they must continually tell everyone how busy they are, particularly when returning from a jolly lunch at 2:45 or after completing some minor project three weeks later than promised.

A more subtle technique involves the proper handling of committees. Improperly handled, committees may occasionally have a useful function. They bring together different viewpoints to assist in finding the optimum course of action and so arrive at a decision. They are also suitable, however, as repositories for projects that you want deferred. Indeed, with proper skill in setting the committee's assignment, there is always the possibility that the committee will never get around to handling that project at all.

Use Committees Imaginatively

There is one elegant use of committees for RA that you may find appropriate, particularly where there is some top executive who loves to avoid personal responsibility by holding subordinates strictly accountable for errors.

Suppose you are a Product Manager, assigned to review specifications for one of your products. You could be foolish and do it yourself. An MAAR, however, would convene a meeting consisting of at least the MAAR's immediate superior, the principal subordinate, a counterpart in at least one other department, that person's superior and principal subordinate, all chaired personally by the MAAR. An imaginative person may be able to do considerably better, but six should be a suitable minimum. Now if the decision is wrong, the responsibility is divided among several people in related functions. An error would have to be serious indeed before even the most hard-nosed executive would cut the heart out of an operation by moving against all those involved!

You, Too, Can Paralyze Your Company

Perhaps the most sophisticated of the Master-rating tech-niques is the following, which, properly handled, can paralyze an entire company. Most decisions, whether by individuals or groups, involve an exploration of alternatives to come up with the course of action that provides as many desirable aspects as possible with a minimum of undesirable aspects. This pro-vides the opportunity for the MAAR to seize upon. Just keep emphasizing the negative. It scares even Eager Beavers out of making decisions. One good way to do it is to bring up extran-eous side issues such as "Do we really dare to market this new product before we test its toxicity on Arctic seals? Never can tell when an Eskimo is going to eat one!"

We now come to those techniques that accomplish all the previous items yet have the distinction of not inconveniencing others by pushing toward them the responsibility you have avoided. Anyone who can do this is entitled to be considered Pretty Hep at Avoiding Responsibility.

The simplest of these advanced techniques is that of the Pseudo-Eager Beaver. Here the individual talks seriously, pref-erably while puffing on an old pipe, about the need for formal decision-making and the techniques of organizing for it. If par-ticularly skilled, the Pseudo-Eager Beaver may even go so far as to set up the mechanisms (being careful not to use them). In-deed, if the Pseudo-Eager Beaver is a true PHAAR, procedures are usually built into the mechanism to cripple the capacity for decision-making, even if in the unlikely instance that someone wanted to make one.

Hiring an Inept Staff

Another elegant technique is that of hiring improper assist-ants. Here the key point is the unspoken question, "What can you expect me to do with a staff like this?" The technique also works with the selection of equipment, by the way.

In effect, you say to your superior, "Do you really want me to lead these green troops into battle, J.B.? Of course none of us minds dying for you, but think of the waste to the organization!"   Once J.B. realizes you can't handle the project with your present staff and equipment, it is easier to abandon it than get reorganized.

There is a special case of diverting attention from essentials that you should learn how to handle. Every now and then a superior finds that there is an Eager Beaver among the subordinates. Since firing someone, even for being an Eager Beaver, is frowned upon in many companies that are more concerned with their public image than the comfort of their employees, it may be necessary to put up with that person temporarily. The situation will eventually arise wherein the EB prepares a study focusing on a decision and submit it to you, putting you on the spot.

A lesser person might be panicked into making a decision one way or the other and hoping for the best. An expert, however, merely picks out some minor item in the report that isn't quite right and sends it back for correction, ignoring the main subject completely! The beauty of this is that while it may take only a short while to make the correction, the subordinate will usually be so discouraged by this turn of events as to decide to forget the whole thing. In fact, if skillfully handled, by including some oblique personal criticism with the rejection, such as a reflection of the preparer's lack of thoroughness, the next and last thing you hear from the Eager Beaver may well be a resignation or at least a request for transfer.

Engineers Can't Communicate—See That You Don't

Inability to communicate effectively is almost universally considered to be a lack of personal skill, and is a characteristic said to be found most often among specialists. It is interesting to speculate whether or not this has a psychological basis as a means of avoiding responsibility. The question of fact can be left for research, but the utilization of the opportunity is available to every candidate for the PHAAR rating.

The technique is simplicity itself; just fail to communicate. When communication cannot be completely avoided, be as unspecific as possible. Since transmittal is only one part of communications, the artist also practices lack of reception. Being inaccessible is a common example. Then when some supervisor wants to know why something wasn't done before a problem got out of hand, you can honestly say "Gee, I never heard a word about it."

The Mark of the Artist

We come at last to techniques that distinguish the true artist. This qualification is easy to describe, but far more difficult to achieve. It requires attainment of full credit for projects that succeed, with absolutely no stain for projects that fail—the epitome of responsibility avoidance.

Analysis will show that this can be achieved only by being on both sides of every question. Thus there is only one basic technique, but fortunately there are endless variations on the theme. The rating is earned, therefore, not so much by recognition of the principle, but by the skill with which it is applied.

Let us consider first the most rudimentary procedure that would qualify for the rating. You are confronted by a situation that requires a decision. It may be a problem presented by an assistant, a report by a comittee, or even an assignment by a superior. You first maneuver someone—anyone—into expressing an opinion about what should be done. This is relatively easy; all you need to do is say, "Well, what would you do in this situation?" in a tone that implies that you are training your subordinates.

Once this opinion is expressed, convene a meeting of some kind. You start off with, "Tom thinks we should do so and so. What do you think?" Give them plenty of time to express themselves, while you listen very attentively. Then lean back thoughtfully and say, "Tom's approach seems very good. Very good. And yet, somehow feel in my bones that it isn't going to work that way. But go ahead, Tom, it's all yours."

That is only half of it, of course. Just as soon as the result is clear, you must remind everyone where you stood on the subject. Recommended statements are, "Well, I not only knew what to do but picked the right man to do it," or alternatively, "I knew in my bones what to do, but you've got to let young people make their own mistakes if you want them to develop!"

Always Delegate

To qualify to the artist degree, the executive must practice subtle over-delegation. Note that the Eager Beaver interacts with the subordinate to whom he or she delegates. The delegator sets objectives and deadlines, and even establishes the criteria of what constitutes success or failure. He or she delegates only what the subordinate can reasonably be expected to handle. The Eager Beaver permits the individual to reach to objectives in any way, but stands ready to help when the person runs into problems.

The true artist, however, uses an approach close enough to be mistaken for the above, but with an important difference. The artistic delegator says, "Tell me what you need to do the job. After I provide it for you, it's your neck if you fail." Note the beauty of this approach compared with the previous one. If it succeeds, the superior obviously is entitled to the credit. If it fails, however, it is the subordinate who failed, for what more could the superior do?

This technique is so clever that its practitioners have even developed an appealing little ritual to go with it. As an example, a Vice President of Marketing selects a Sales Manager and delegates to that person full responsibility for company sales, taking care to keep aside several large accounts that the VP will continue to handle personally. With the steady growth of the chemical industry, things go well for a while and the VP is a hero. Eventually, problems develop, possibly because of a recession in the national economy. The President or Board of Directors asks the VP what is wrong and what is being done about it. Now here is the ritual you must perform and the magic words, which you must memorize: "I put my faith in Joe, and he let me down." So saying, you take your magic knife and remove Joe's head. Then you find another Joe or Jane. Even if suspicion develops after three or four repetitions, you still have those little old accounts!

In this short course it has been possible to touch on only a few typical techniques for responsibility avoidance. The techniques are limited only by the ingenuity of the practitioner, and new ones are being invented every day. Much better than a better mousetrap, really, as a route to success.

Carl Pacifico
Chemical Engineering
P. 149, November 20, 1967

Anthropleaogy

Won't you please finance my expedition
To discover yet another missing link
It lies between Australo and the Homo
And can be found in Paris France, or so I think
I intend to deeply study all the primates
And from this conclude where the fossil lays
I will study them through none but finest crystal
I will study them in little-known cafes
All that stops this excavation is some money
Not ignorance or thoughtlessness or fear
I will dig him up with very ginger fingers
And name him *Francaispithicus* (add name here)
Consider the import of my proposal
Your name rung out in every lecture hall
Printed in italics in each journal
And scratched in paint on every bathroom wall
My search will be performed with such integrity
I will dig from soil to bedrock down to hell
And after, with the help of your donations
I will quest for Francais Femalae just as well.

Contributed by

John Kearney

Finding the Meeting

Dr. Karl K. Darrow, long time Secretary of the American Physical Society, was an incisive critic of speakers in almost every area of physics, a stylist (*The Epitome of the Programme*), and a meticulous planner of APS meetings.   (He advised those submitting papers that in his experience white envelopes traveled faster than manila even when both were sent by First Class mail.)

Dr. Darrow's explicit instructions on how to reach an APS meeting were much appreciated. The following example was not only reprinted several times in the *Bulletin* but also in the *New Yorker Magazine*.

How to Reach Columbia University

To go from the Hotel Pennsylvania to the Pupin Laboratories, proceed to the IRT subway (under the front entrance of the hotel), and take an uptown train marked either "B'way-7th Ave Express—Van Cortlandt Park" or "B'way-7th Ave Local —137th St." Get off at 116th Street, and walk three blocks north to 119th Street, turning there to the right through the campus gates. Allow 35 minutes. If on emerging from the subway at 116th Street you find yourself on Lenox Avenue, you have failed to follow the foregoing instructions, and should take a train in the reverse direction back to 96th Street again (it would, however, be quicker to take a taxi).

Karl K. Darrow
Bulletin of the American Physical Society
Vol. 19 (1), Jan. 14, 1944

Views Across the Disciplinary Fence

The Compleat Engineer

Chemical engineers are those who know...

Enough engineering to confuse the chemists
Enough chemistry to confuse the draftsmen
Enough mathematics to confuse the boss
And enough electricity to confuse themselves.

Their world consists of a series of simplifying
assumptions and derived basic equations
All of which they are sure are true
Because they can be integrated twice.

Contributed by
Richard H. Okenfuss

Theorists

To some minds, geology scarcely assumes the rank of a science, except when treated from a physical point of view. They consider the simplest physical law adequate to the explanation of the most stupendous phenomena. To them, mountain chains rise and are abraded, and the entire crust of the earth is folded and plicated in obedience to certain laws. They see no difficulty in the way of imagining torrents of water moving onward and upward, carrying masses of rocks over

heights far above their origin by some simply gyratory force.
The entire earth becomes with equal ease to them, either a pli-
able, elastic, or compressible mass, or a nonelastic body; sometimes one, sometimes the other, changing gradually, or by sudden cataclysms, according to the fancy of the expounder. No
wizard's wand ever played so many pranks as the poor earth in
the hands of these theorizing geologists.

J. Hall
*Contributions to the Geological
History of the American Continent,* 1882

Medicarnage

Cogito Ergo Sum

If I accept as fact, not sham
Descartes' "I think, therefore I am"
It follows that of my friends, a lot
Who cannot think at all, are not.

Analysis

Freud
Toyed
With the lid
To the id.

The Psychoanalyst

This study of your mind and sex
Gives me the facts, Mister Dior.
You suffer from no dark complex.
You simply are inferior.

The Gynecologist

Ad nauseam
Per vaginium

The Bacterium

Bacterium is just a term
Pathologists use for a germ.
Some are round and globular,
Some are rods and regular.
In size they are mixed and various.
Their intentions generally nefarious
But though we think of germs as pathogens,
Some are really our best friends
But for me it is quite a feat
I just don't feel for the spirochete.

The Status Stigmata

My face is wreathed in beatific smiles
Because my other end does not have piles.
If fate decrees I must be sick,
I'd like it to sound exotic.
I'd hate to have to tell my friends
I'm suffering from the bends.
I find it degrading to be told
I've merely got a common cold,
While a thing like Asiatic flu
Makes me one of the privileged few.

Ian Rose
from *Medicarnage*

The Medical Archives of Rattionalia

Accidentally separated from the jet plane that was carrying him home from the International Congress of Tonsilology, Dr. Lemuel Gulliver VIII landed by heredofamilial good fortune in Rattionalia, the country of the laboratory rat.

During the first hours after his arrival, he gave himself up to impending death. He was told that human beings had been used as laboratory subjects by early Rattionalian investigators: They had been dispensed with only because their nervous and psychic responses lagged behind those of rational animals, and because they refused to breed true. Dr. Gulliver's dexterity with the laryngoscope (he had thought to excite his captors' awe by exhibiting that instrument) and the evident purity of his lineage roused the avidity of the scientists to whom he had been consigned. Only when he chanced to mention that he was the author of numerous publications in his field, and an associate editor of *Throat,* was his demise postponed.

"In that case," the neuropsychiatrist who was examining him reluctantly conceded, "you are in Dr. Whisk's jurisdiction."

"And who is Dr. Whisk?" asked Gulliver bravely.

"Dr. Whisk is Chief of the Word. It is he who decides the destiny of medical authors."

On the morning of February 30, Dr. Gulliver mounted the plastic ramparts to the office of the President of the National Scientific Archives with profound apprehension. He knocked and was admitted.

Dr. Whisk was a white-whiskered rodent of calm visage, the product of 19,382 generations of inbreeding in the D-86Ar2 strain, every centimeter an aristoc rat. Appearing to defer the tasks of justice, he implied that he was Gulliver's host. Within minutes, Gulliver was disarmed, not only by Dr. Whisk's benign manner, but also by the splendor of what he saw. The archivist escorted him, through a wall-to-wall carpet maze, to the National Library of Medical Publications, explaining as they went that each metropolitan district, together with its adjacent rural area, was served by a duplicate of this institution.

Despite a wealth of evidence of a rarefied order of knowledge, the national medical library was contained within a single room. Emblazoned above its entrance was the motto, MISERERE, RATTE, SILVARUM. About a hundred librarians answered hushed telephones, sought volumes in stacks, and placed them opened under Xerox machines. Dr. Gulliver looked in vain for a reader's room.

"No doctors?" he asked.

"Heavens, no! We couldn't afford interruptions, the chance of loss, damage . . ." Dr. Whisk nodded toward the switchboard. "Every physician and surgeon in Rattionalia A has a direct wire from the office to that board. In our technical journals, each published article is accompanied by a card of standard size and format, on which the article is summarized. The card is both punched for electronic sorting and printed with a code for hand filing. In the national libraries the cards are filed electronically; most physicians, in their offices, use the hand system. For a nominal fee, one may subscribe to the summary cards alone. Within minutes after physicians call this room, they receive by direct wirephoto a facsimile of the article requested. They may have copies of the entire literature on a given subject within a half-day."

"I read once," Gulliver commented sceptically, "that the cost of a system like this for a country with the scientific output of the United States of America would be prohibitive. I suppose for a small nation like yours...."

"The source of this disparity," the rat assured him quietly, "is not the relative sizes of our countries." He added with a dry smile: "The administrators were somewhat recalcitrant about the initial cost; but the maintenance expense is a fraction of that drained into dozens of inadequate libraries in hospitals and medical society offices under the old system. But tell me," he asked, "do your doctors still drudge from office to library, from index to index, in medieval acquiescence to senseless labor?"

Ignoring the query, Gulliver sauntered about. His hand itched to remove a volume from the shelves, but he had been warned not to touch the books. Dr. Whisk beckoned to a librarian, who sped to their side and with barely masked reluctance handed them a four-year old copy of the *Journal of the Rattionalian Medical Society*. The first article began and ended abruptly:

Hwvr, stxla appr to be snthszd in ncls of lambda cls.
Rtio cytplsm:ncls consistntly l:l. Spcfc actvty
Stxrla^{911} in cytplsm excds that in cl wll; hnc cytplsm
stxrla not from cl wll. <u>Method:</u> Stxrla^{911}, M862Kt.

Dr. Gulliver blinked. "That's the article?"

"An important contribution in its time,"   Dr. Whisk
assured him.

"A bit succinct, isn't it?"

"It was all he had to say that hadn't been published
before." The rat inclined his head reverently toward the motto
above the doorway.

Back in Dr. Whisk's office, Gulliver seated himself on the
floor. Loyalty to home and species, and to his somewhat hum-
ble subspecialty, the human tonsil, forbade that he confess his
uneasiness. Dr. Whisk seemed to discern the reason for his
silence.

"To know a field well," he told Gulliver gently, "one need
not rely on endless repetition.   Indeed, when the reasoning
psyche is assailed by repetition, it reacts with doubt. The most
efficient way to fix a fact in memory is to see and hear it demonstrated once, adequately, within its context; then to apply it.

"Our literature," he went on, "is a formalized conversation
between investigators and clinicians. Each of our textbooks is
a compendious review including three facets of the given subject: the totality of proved data, the comments of all authorities
in the field on moot points and the questions to which answers
are currently sought. These are our baseline publications. Man-
uals teaching their use are employed in the medical schools.
One text in each specialty and subspecialty is issued every five
years. During the interim, none may appear.

"Our technical journals are edited by the same bodies of
rats that edit the texts. Rotation of editors from clinical and
investigative fields assures both freshness of approach and
continuity of method. The total literature may be read as a single unified article."

Gulliver recalled the awe in which Dr. Whisk was held.
"You originated this idea, I suppose?"

"By no means single-handedly. It evolved as a product of
many minds. Although three-quarters of our editors are special-

ists in clinical and investigative fields, one-quarter are physicians specializing in medical communication. Being as zealous for advance in their skill as are radiologists or surgeons, these specialists constantly improve the means of communication."

"I suppose all this terseness and smoothness and so on speed your investigative work?"

"Considerably."

Gulliver's mind raced to encompass a vision of illimitable acceleration. "How do you, as publisher, keep up the pace?"

"When material is presented to the editor in this style, publication is simple and swift. Underlying the product that you see, Dr. Gulliver, is a basic principle: In Rattionalia, the process of assimilation is demanded primarily of the author. It is capacity for assimilation that makes authorship kin to authority. Once that responsibility is accepted by the writer, the editor's task is as simple as his role is self-effacing." Indulging his enthusiasm, Dr. Whisk added, "These principles serve us well. You, for example, although a newcomer, could learn from our literature all that we know about the tonsil within a month."

Gulliver got up and prowled about. The ceiling was lofty by Rattionalian standards, but he found it oppressively low. "The whole thing's too damned inhuman," he grated. "I'd miss the old, comfortable puddle. You can say all you like about facts *versus* opinion, but when I see a necrotic tonsil, I want to know more about what a first-rate doctor thinks than about what some teenager in a lab knows."

The first-rate Rattionalian nodded. "We have, of course, another type of publication—'whipped cream,' we call such —wherein writers may divest themselves of their emotional loads and the reader may obtain vicariously a similar relief. We have tape recordings, too; closed-circuit televisions, motion pictures, verbatim transcripts of bull sessions—all these have a certain value. But we do not confuse them with scientific reports, and we do not index them."

Gullliver hunched on the carpet to ease his neck, and scowled. "How do I know that the author of that article used proper controls?" he challenged.

"An excellent point." Dr. Whisk placed the tips of his paws together in a gesture used by pedants the world over.

"We decided decades ago that ascertaining the reliability of an investigation is the function not of the reader of a scientific report (who cannot absolutely determine this in any event), but of the institution under whose auspices the work is done. Colleagues from the author's university or hospital scrutinize his methods and results. A board of specialists from another institution, selected by an electronic straw-drawer that practically eliminates the probability of bias, reviews each investigator's work. Thus, as far as rattily possible, we have evidence of validity before committing a statement to print."

Dr. Whisk punctured his epidermis lightly and brushed on a drop of tranquilizer–stimulator with a gentle, stroking massage—a civilized equivalent of smoking that Gulliver had considered but had rejected because it might form a habit he would be unable either to continue or to forget after his return.

"How about the medical schools?" he asked suddenly. "Twelve authors to a paper—papers broken into subsections to make multiple publications—how do you help academic aspirants satisfy faculty requirements?"

"The medical schools have been our greatest source of help. Publication remains one criterion for promotion, but other criteria are equally valid. For example, one of our best proofs of our investigators' acumen is their refusal to publish results if they doubt their importance, no matter how much time they may have spent on obtaining them."

"Well, now, it's the same with us—"

"When did you last give an official dinner—and a raise in pay—to an investigator who buried a year's intensive work in silence?

"Silence?"

The rat regarded him with lightly veiled amusement.

Gulliver rested his head on his hand and sighed.

"All right," he conceded at last. "You've got something. Now, how did you bring it about?"

Dr. Whisk's black eyes twinkled reminiscently. "It all happened abruptly in the end. Back in the years 1950–1960 A.R., all but the most bedazzled of us realized that we were deep in an epidemic of the moronic plague. Our first thought was to set up a Research Program in Technical Communications, to cost the appropriate millions" He smiled leniently and eased back in his chair. "We appropriated the money, formed a committee (I was a junior member), and started the pilot study. We had the enthusiastic backing of every scientist in Rattionalia. "Then, he glanced quizzically at Gulliver, "our committee discovered that its first step must be to psychoanalyze every author who had published a paper within the last ten years to learn hidden motivations to publication." He chuckled softly, "Now, of course the most influential rats were those that had published most. Within days, laws were passed against unnecessary verbiage in technical literature. Repetition was voted a crime. The punishment is fiendish but effective."

The black eyes twinkled again, and Gulliver's blue ones began to emulate them, but the twinkle hardened to a glint and Gulliver's smile slacked. The memory of an extensive review he had submitted within the last week to the *American Tonsil* rose to the surface of his consciousness.

"A fine, I suppose," he murmured casually.

"Imprisonment."

"You're mad! You put a fellow in jail for—"

"Not in jail, Dr. Gulliver. We consign the repetitive writer to the medical archives of the days before the reformation. Do let me show you."

Dr. Whisk rose and conducted his captive guest to a waiting elevator. It carried them down interminably; they stepped out into blackness; and the elevator rose behind them with a hushed sigh. Dr. Whisk touched a switch. An ultraviolet light spead a ghastly color over endless stacks of bound volumes receding into murky catacombs. The musty smell of fine leather invaded the nostrils. Gulliver touched a book and his hand recoiled. The thing was bound in human skin.

"Take down a volume," urged Dr. Whisk. "Glance through it. The old style has a certain nostaglic charm."

Stifling his loathing, Gulliver opened the first book on which his hand came, not to rest, but to quiver. An introductory paragraph looked as familiar as a scuffed doormat:

As early as 1873, it was noted that the bacterium *Styphaureomalcides botuparagraphylii* was to be found in...However, it was not until 1911 that Murchison discovered....Our interest in this field was aroused when we observed by chance that...

Gulliver slammed the book shut. "You can't do this too me!" His scream rasped through the catacombs.

Dr. Whisk's teeth gleamed in the hideous light. "No cause for alarm," he assured Gulliver silkily. "You're not guilty, I'm sure, of such a crime as this?"

"I swear I never—" The words struck in Gulliver's throat. "Look here, I'm not punishable by you! I'm not a citizen of your filthy rat's nest—"

"And therefore have not a citizen's right to a court trial," grinned the rodent. "All that you have unwittingly revealed by your crude astonishment at our methods...your attitude of self-defense at this moment...these and more suggest—"

He crowded against Gulliver. The human physician overcame his revulsion against the rat by dint of his greater detestation of cowardice.

"There are things you've forgotten," he urged with forced calm. "Like being friendly to strangers. The love of humanity."

At the word "humanity," the rat turned away and gave three short taps to the elevator button. When the car arrived it was full of rats in uniform. Each carried a weapon the size of a fountain-pen, which looked mightier than the sword. Dr. Whisk joined them with dignity and turned to face his prisoner.

"In this rational country we don't extend punishment beyond its remedial effect." He smoothed his white whiskers with a benign smile, more like a pediatrician prescribing a tonsillectomy to a bewildered child than like the ruthless dictator of the Word. "I ask you, sir, as my guest, to spend a day and a night here. Food will be brought to you. And on the shelves, there is so much food for thought. Make yourself at home; you'll find the atmosphere familiar."

The elevator door closed. In the eerily lit archives, Gulliver stood with the rat's last words whispering through his skull. "Just spend your time reading, Dr. Gulliver. Reading endless repetition. Endless repetition. Just Read. Read. Read..."

Ruth Straus
The Lancet
p. 1182, December 26, 1959

The Difference Between Chemists and Physicists

I have always been interested in the difference between physicists and chemists. It is possible to understand chemists—that is, most chemists—but physicists are different. My difficulty in assigning physicists to their proper place in the cosmos may possibly arise from my gaining acquaintance with chemistry earlier than with physics. Having had the good fortune to be brought up in an old house full of books, among which were some dealing with chemistry, I early gained an idea of the aims and objects of that science. There were no books dealing with physics, which was fortunate. I can think of no more evil influence on children's lives than to let them read the works of physicists. Losing all regard for facts, they would grow up with the loose and faulty reasoning so characteristic of clergymen, while lacking any of their saving charity and comfortable optimism. They might even come to believe in the atomic theory.

At college I very soon found out what physics was about. It seemed that our professor had had it revealed to him by some kind of super-professor who had learned it in turn from a Professor-Extraordinary or a Geheimrat or something of the kind in Germany, that the world was made up—every bit of it—of very small particles in very rapid motion. The exact details of this motion had been arranged a long time previous by a Mr. Boltzmann, and the size and shape of them by Lord Kelvin and Mr. Rutherford, who used to teach at McGill College, and a Mr. Bohr—some kind of Czech or Russian —and a lot of other great men. These things were all settled, and there was no way of changing them, no matter how many facts emerged that seemed to make it advisable. It was like the American Constitution.

Many years after leaving college, my awareness of the differences had only grown more sharp. Thus, I now saw that a physicist generally does one or two experiments before writing a book of 1734 pages, whereas a chemist typically does 1734 experiments and writes up only a page or so in a notebook about them. There are other kinds of chemists who don't make experiments at all. They take out patents.

I have also learned that it is sufficient unto a glorious career for a physicist to know calculus. Calculus is to many physicists what the Constitution is to a lawyer. After having differentiated the results of an experiment transposing them and integrating them once again, a competent physicist will have a new experiment and several hundred pages toward the next book.

There are several sorts of physicists. Old-fashioned physicists are merely high-church engineers. Today, however, there is a sort of Greenwich Village physicist who is enough to drive anyone insane. I have just read a book about one of them. This book proves that "2" is a number. It is very convincing if you have ever had any doubts on the subject. These men are very much interested in what they call the "Principle of Relativity." This means that a pound of lead up in Central Park at midnight isn't necessarily the same as a pound of lead down at the corner of Chambers Street and Broadway the next day at noon. Of course, you couldn't prove that it was different—no fairminded scientist would ask that—but, on the other hand, you can't prove that it only seems the same because of experimental error. Some years ago a pioneer of this school in Pennsylvania discovered that Alice in Wonderland was really a work on hydrodynamics written in cipher. This sort of thought is considered very advanced by some, but the authorities locked the adventurous cogitator up in the Butler County Asylum nonetheless.

It is probably a good thing to have men of this class. Nowadays we have "free" poetry and "free" art and "free" music, and so, why not have "free" science? One can hardly expect the younger modern intellectuals to be bound by such old victorian conceptions as gravity or to hold closely to the idea that $PV = nRT$. However, most of these "new" physicists are rather a bad lot. They are loose in their morals, are frequently pacifists, admire music by Philip Glass and poetry by Allen Ginsberg, and as likely as not they are pro-Khmer Rouge.

Physicists differ from chemists in not being able to earn their living. Most professors of physics marry rich wives or have friends who tell them what stocks to buy; or they have daughters in the movies. If not, they die in the workhouse. I am glad I am a chemist.

F M T, Jr.
The Percolator
The Chemist's Club of New York, 1918

Information and the Ecology of Scholars

What is that very large body with hundreds and hundreds of legs
*   moving across the horizon from left to right in a steady,*
*   carefully considered line?"*
"That is the the tenured faculty crossing to the other shore on the
*   plane of the feasible."*
"And this tentacle here of the underwater Life Sciences Department..."
"That is not a tentacle, but the department itself."

Donald Barthelme, in *"Brain Damage"*

As the behavior of biological ecosystems becomes increasingly familiar, one is tempted to use ecological phenomena as metaphors for human behavior. The apt and often amusing insights so produced have seemed, however, to offer little basis for the serious consideration of quasi-thermodynamical systems models for society. My object is to suggest a logical basis for the use of ecological concepts in modeling a special subculture: that of scholars (and, in particular, scientists), who produce, barter, and structure information as an ecosystem produces, exchanges, and structures biomass. To go beyond mere analogy, however suggestive, it will be necessary to discuss ecosystems as open, dissipative thermodynamic systems and to point out the relationships between material (thermodynamic) and conceptual (informational) structuring.

I begin with the common observation that the orderly structures found everywhere in nature may be classified into two groups: equilibrium structures, epitomized by crystals, and dissipative structures, epitomized by living cells. Both kinds of order represent relatively low-entropy states for the matter of which they are composed, although both may persist unchanged for long times.

Dissipative structures depend for their persistence on a flow of energy; if this energy flow is cut off, the structured matter begins immediately to relax toward a state of equilibrium, in accord with the second law of thermodynamics for a closed system. Low-entropy states of matter may also, through information theory and statistical mechanics, be characterized as states of high information content and of low probability. I use all four expressions (order, low entropy, high information, low probability) as equivalent, bearing in mind that the existence of quantitative correlations among the four allows such interchanges to be made in fact for particularly simple situations (like ideal gases, telegraph messages, or statistical distributions among a-priori equally-probable states) and in principle for complex ones (solutions in condensed phases, genetic and behavioral information, or social organization).

The Social Ecology of Academe

There is a formal resemblance between the community of scholars and what biological ecologists have been willing to call ecosystems.   Consider the definition of ecosystem: a group of living organisms related by their common access to a well-defined energy flow and by their participation in a web of nutrient and informational flows. No problem should present itself in catergorizing scholars under this definition, if it is remembered that the supportive energy flow includes the flow of information according to the hierarchical structure discussed above, and what distinguishes the academic ecosystem from other units of the fossil fuel economy is just this additonal peculiarity.

A consideration of the four properties of dissipative structures, as exhibited by ecosystems, should serve to test this conclusion.

(1) *Dependence on energy (and information) flow*. It is hard to see that more needs to be said here. The physical structures of academe require maintenance energy in the form of electrical power, physical work, and money. The social structures require, all too clearly in this era of academic under-employment, money, but also psychological motivation.   A man becomes a biologist, say, and a member of a biology department. He works to make both roles satisfying, because to do so earns him a salary and because he wants to understand biology. Only the intellectual structures survive on motivational energy. (One would, of course, be tempted to distrust such an opaque term as "motivational energy" except for the fact that this is a quasithermodynamic, not a mechanistic, model. Lacking human motivation, the intellectual structure ceases to grow and ultimately disappears, and that fact is sufficient to identify it as a dissipative structure and to validate the term "motivational energy.")

(2) *Homeostasis.*   Academe has survived war, famine, plague, Christianity, prosperity, and, so far, the SDS (Students for a Democratic Society) and the FBI (Federal Bureau of Investigation) with remarkably little change in structure or method for over 2000 years in the West. This stability probably results in large part from a strong historical tradition (embodied, for example, in the institution of the textbook), as well as from the fact that the human mind appears to be constrained to build constructs according to a nearly invariant logic, using models derived from daily experience.

(3) *Succession.*   Within this nearly constant framework just mentioned, there occur cycles (both on a local scale and, on a few occasions, encompassing the whole system) of greater or lesser productivity. The dynamics of these cycles have attracted the attention of sociologists, and a few common features have been pointed out. Generally, a period of high scholarly production follows the advent of an entirely new model—in Kuhn's term, a new paradigm—that serves as the basis for a new way of looking at the world. In terms of the hierarchical model of scholarship, a new paradigm appears high in the hierarchy, and much subsequent work is devoted to assisting the consequences of this change to propagate downward through the logical structure. A biological system would experience a similar effect from a sudden pulse of nutrients or other form of energy. The response of the biological system is to return to a less mature state. It has been my observation, and it is documented in a more scholarly way by Griffith and Mullins, that groups of scholars react to a new paradigm in ways that, according to the ecosystem model, can be characterized as a return to (ecosystem) immaturity. Lines of communication become short and rapid; motivational and fossil fuel energy flow are high; diversity within the active research group decreases, as all members share enthusiasm for the new paradigm. Naturally, such a state cannot be maintained for long. As the group grows, administrative structures tend to grow with it, energy is diverted to maintain the larger structure, paperwork multiplies. At the same time, ties to the larger world of scholarship are reestablished.   The informational pulse, in short, is metabolized and elaborated into a complex and flourishing structure, in which information webs grow up to replace

the simple, linear pattterns of direct (and excited) oral communication.

(4) *Kinetic limitation of growth.* The economy at large is limited in its growth and productivity by material scarcities; the academic ecosystem, tied as it is to expensive facilities and mortal men, shares this limiting factor. A much more crucial and interesting factor is that academic systems are also limited by the rate at which the consequences of new paradigms can be worked out. In biological systems, kinetically regulated nutrient scarcity gives a competitive advantage to specialized species, which helps to account for the increase in species diversity as the amount of biomass grows toward the limit that the available fluxes of sunshine and nutrients will support. In academic ecosystems, kinetically caused information scarcity gives a competitive advantage to intellectual specialists. After all, there would be no advantage to specialization if the logical consequence of any paradigm or the optimum use of any technique could be perceived quickly by anyone. It is common experience that the longer an academic field matures, the more intense and diversified the degree of specialization within its ranks. It seems perfectly consistent to attribute this successional phenomenon to the slowness with which the human mind produces new paradigms and works out their consequences; that is, to an ultimately materially determined limit to the rate of nutrient flow in the academic ecosystem. The high structuring, the scarcity of information, the slowness of thought produces niches (to use an ecological word for the phenomenon) that are remarkably constant from one field to another: artist, critic, graduate student, archivist, bench worker, the seemingly inevitable handy folks who are assigned to the care of audiovisual equipment. Such niches tend to be shifting or absent in the small and democratic "hot" groups that first gather around the personal or geographic center of a new paradigm .

Research Institutions

A research institution is in the business of producing and structuring information, as a farm or forest is in the business of producing and structuring nutrients, given certain energetic and material throughputs. It appears that the productivity of a research institution is not governed solely by the size of those throughputs (in spite of the necessity of *some* throughputs and the tendency of funding agencies to reward large information outputs *a posteriori*). With increasing maturity, biological ecosystems tend increasingly to substitute behavioral and physiological controls on their production rate for physicochemical ones. Perhaps the same is true of academic ecosystsems. If that should be the case, an interesting dilemma arises: it is in the direct interest of the personnel of such an institute that the productivity be as high as possible—that is, that the institution resemble Giffith and Mullins' "coherent groups." On the other hand, there is a real human tendency, especially in large groups, to devote physical and mental energy to the creation and maintenance of more and more intellectual and organizational structure. No wonder, then, that the progress of science shows alternations of structure-breaking new paradigms with structure-building quiescence. It may not be too farfetched to suppose that the tendency to become increasingly committed to the known complexity of a "developed" paradigm is a result of humanity's psychological fittedness for life as an integral member of a climax forest community.

In any event, the indulgence of this drive works against the interest of the institution as an economic entity, and a good director will seek to disrupt too comfortable an ambience (for example, with nutrient pulses in the form of provocative seminar speakers or by discouraging the creation of divisional boundaries within the institution) and to push in the direction of (ecosystem) immaturity. Evidently, a question of optimum strategy arises, in which the free productivity of immaturity is balanced against a psychological preference for the comfortable complexity of maturity. It may be possible to account for the differences among institutions by referring to how successfully they have solved this dilemma.

Thomas R. Blackburn
Condensed from *Science*, 181, pp. 1141–1146
September 21, 1973

The Chemist

The Analytical Chemist

Most accurate of chemists, this professor
Is the world's most prolific second-guesser,
Who from two wide extremes may fuss and fiddle
And end up with an answer in the middle.

Upon worn thumb and forefinger the blister
Of the pinchclamp pincher and the stopcock twister,
The analyticker studiously and concisely will
Determine chlorine in a chlorophyll,
And may the fate of future years foretell
By thumbing through old Kolthoff and Sandell,
Or may detect, if your request be louder,
Impurities in Mrs. Murphy's chowder,
Who finds for you in manner unperturbed,
The arsenic your wife slipped in the sherbet.
Most versatile, this calibrator of many talents
Thinks anything is possible with a balance,
But will confide to any trusting brother
That life's just one damned pH after another.

The Inorganic Chemist

Most chemists are unwilling to say if
They have ever heard of comrade Mandelejeff,
Or to answer with no evidence of terror
Whether or not rare-earths are getting rarer.
But the inorganic chemist will merely snicker,
And point to freshman textbooks, getting thicker,
Or scowl if you should mention old Arrhenius,
And take off on a list of newer manias.
He has led old war-horse Bronsted to the stable,
The inorganicker has massacred the Periodic Table,
But still fills tubs with brine and pins high hopes
On growing crystals big as canteloupes

One causes considerable ruction
Over which is oxidation, which reduction,
And seems ever competent to, more or less
Fill up our breathing-space with aitch-tu-ess,
The inorganicker's the only purist,
And any other chemist's just a tourist!

The Biochemist

These are the extractors of the clan
Who pry from sundry garbage what they can,
Who drop the well-fed mouse into a vat
And make a savory soup of fur and fat,
Who wield a shovel in sanitary pens,
Or steal the eggs from calculated hens,
Who feed unknowing microbes chosen dinners
Then chart the habits of the little sinners.
These are the scientists who have a way
Of getting chemistry from baled hay,
Or render from stewed cornhusks smelly messes,
The uses of which are anybody's guesses.
Who give well-chosen headaches to a steer,
Or track the course and disposal of a beer.
These characters are very certain and emphatic
That hormones make you sexy or rheumatic,
And always get eight hundred compounds—very pure—
From twenty thousand tons of good manure.

The Organic Chemist

Borer of corks, the skilled Organischer,
Is interested in all things that stir,
And will urgently, in manner most implausible,
Get as many fingers into the flask as possible.
And if the result is some scorched solid,
Our doughty cook will fracture it,
Along the flask that helped to manufacture it.
Sublime, the great organic chemist basks

In a world of broken beakers and dirty flasks.
Heaven to this heedless chef is some Elysian field
Where every reaction gives a perfect yield.
The curse of goaty odors dogs the organicker's toil
Where nothing can be stirred and all will surely overboil,
And soon become apparent without impunity
To the nostrils of the whole community.
Marked by hard labor, this alchemist has a hand of gold
(Of highly nitrogenous content, so we're told)
And one would think that any cook were surely daft
Who would sit up all night with an ailing Friedel-Craft,
But old organicker will tell you in manner quite bombastic,
What fails as pure can most certainly serve as a virgin plastic!

The Physical Chemist

Here is a species, nay, almost a phylum
Of chemist who has gained to high asylum
And achieved a most involved and envied status
By calculus and all-glass-apparatus.
If measurement is king, then prince is theory,
But of each member's view, the rest are leary
And oft' a Physikalischer's life is spent
In writing up that lifelong lone experiment.
To fill the literature with weighty patter
On the vagaries of one small bit of matter
Must be the single obligation and obsession
For the lower-minded of our great profession,
Who can tell you to three decimals on a slipstick,
The density of Maggie Thatcher's lipstick,
And upon some measurement of surface tension
May hang the fact that gives us all a pension.
Psychologists of atom-life are they
Who cut their eyeteeth on $PV = nRT$ and the gamma-ray!

The Chemical Engineer

Let us not grieve that modern times installs
Its progress in a pair of overalls,
Nor that long training gives a muscled wrist
 By which important valves are given twist,
Nor that long lines of endless pipes must totter
And expend themselves to boil a pail of water.
We should not deem that progress is dismayed
By Graham's law unimpressed or disobeyed,
By columns towering sleekly toward the blue
With theoretical plates of minus two,
But if from such adversity we flinch,
The ape-that-walks-on-two-legs-with-a-wrench
Will still link solvent line to manostat,
And filter-press to mother-liquor vat,
And peer at many-needled meterata
With which to fill an endless book of data,
And better things for better life will rise
Triumphant from these views on chemistry,
Because no matter what the slide-rule rules,
Molecules will still be molecules!

Chemist Retired

"What was your place in Life?" He asked,
As I crossed the Glorious Portal.
"I was a Chemist, weird and impulsive,"
I said, "but essentially mortal!
Probing within at the basis of Sin,
Not outwardly at the Infinite,
Piddling with Pills for a few human ills,
And lighting some candles at twilight!"

"What did you do then of worth?" He went on,
As I stood at the Threshold Sublime.
"I spun Golden Thread for the duPont clan
And taught them to use it in Time!
Working in rooms with innumerable fumes,

With profound expertise and emotion,
Under Servitors vain who were quick to complain
And imbued with incredible Caution!"

"Why did you do it?" His next question was,
As I paused at the Glamorous Gate,
"My family was starved for the good things of Life,
And I learned about Heaven too late!
It was not for a Fact that the Molecules lacked,
And I worked not for Science, but Money,
And that was the reason!" I said, and He smiled,
As though I had said something Funny.

He didn't say "yes" and He didn't say "no,"
To my ultimate Spirit address,
And since chemists work poorer in Heaven than out
There really was no chance to Confess.
I have found as a Ghoul there is plenty of Fuel
In the most likely spot you could name,
A most diligent Search showed duPont and the Church
Were administratively quite the same!

Read by
H. A. Hoffman, DuPont
Chemist, at his farewell party in
Camden, South Carolina, 1980
Updated by the publisher

BOOK REVIEWS

Two Typical Books from American Psychologists

Book reviews abound on the scene of American psychology. They are found in the learned journals, on the jacket blurbs of publishers, and in not a few undergraduate term papers. As such they represent an important part of psychology, and are fair game for the social psychologist.

Below are two book reports that the writer feels represent two major extreme positions in American psychology. Both purport to be reviews of books by critics who are themselves believers in the position taken by the author. Both are fictitious. Considered together, they represent, in my opinion, the sum total of perhaps half to three-quarters of the reviews of books published in psychology (exclusive of the work presenting data on sensory organs and the like.) And both are Horrible Examples of that peculiar narrow-mindedness that attempts to define a science too closely.

In the Bowels of the Soul

by **Wilhelm Oenkschwein Ritter von Kluernk**

From the bulging storehouse of his 20,000 hours behind the couch, this wise old lover of suffering humanity has deigned to present his deep insights into the sick, organismic unity that is the multicolored, ongoing phenomenon of the dynamic, warm, juicy, oozing human personality. Eschewing the use of so-called "data" by so-called "scientific" psychologists, Dr. von Kluernk peers deep into the wellsprings of psychobiologicoanthropologicochemical evolution, underlining the transcendental imminence of deep urges as made manifest in the profundity of unconscious symbolism. No vulgar data or coarse statistics mar this exquisitely delicate work; pagination, for example, is accomplished by the use of mystico-erotic symbols from medieval books of Black Magic.

Dr. W. O. R. von Kluernk's theory of personality (the "neo-Oceanic" theory) holds that the Basic Human Urge is a deep instinct to return to the warm salt water seas from which our ancestors emerged. People make money to return to the sea (symbolized by Miami Beach) or are overcome by the nasty old inhibitions of our rotten, mechanistic, sanitary society and move inland in a masochistic reaction formation (symbolized by Kansas).

The author's achievement is especially dramatic when one considers that he has not left his office in 40 years, during which time the only humans he has seen have been his patients.

Only through such shattering empathy can the full richness of the ongoing organismic dynamicism of the wholistic human personality be comprehended.

Notes Toward a Bare Outline of the Empirical Operational Behavioral System

by Calvin Mather Brittlesparse

This book is the most rigorous statement yet from the rockbound theorists of the Ivy League colleges. It is scientific and objective and thoroughgoing and clean and hard and true. It stands firm on a bedrock of purist operationism. It is unspoiled by speculation. It is unspoiled by thought. The author wants only data. Here are the data. On them the author builds a science that is pure and straight. Now the physicists will not get more money and chairmanships of committees than we.

The completely controlled experiments described in detail begin with the invention of the Brittlesparse box. It has no lever. It has no food. It has no subject. To operate it, the box is hit. It breaks. By thus eliminating the organism completely, a parsimonious science of Stimulus (hitting box) and Response (breaking box) is achieved. The goal of psychology has now been attained. Complete control and prediction is possible. If you hit the box it breaks. If you do not, it does not. This is it.

Summary

Reviews of two imaginary books in psychology are presented as a satirical comment on extreme positions and spineless reviewers.

Stanley A. Rudin
Psychological Reports
Vol. 5, p. 113, 1959

Classroom Emanations

The Way It Used To Be in Chemical Education

The current furor over education in science prompted M. C. Usselman of London, Ont., Canada, to pass along the chemistry syllabi at the University of Edinburgh for 1866–1867 and 1873–1874. The students must have suffered frightfully. Usselman also sent in the doctoral examination in inorganic chemistry in 1872. The written part occupied two days and went like this:

Write a paper on *one* of the following subjects:

1. Classification of the elements. Various systems of classification. Illustrations and criticism.
2. Laws of chemical combination. Statement of experimental basis. Empirical laws derived therefrom. Discussion of the theories by which they have been explained.
3. General account of the most important metallurgical processes, with full discussion of the chemical actions taking place in each.

*   *   *

Doctoral candidates also were given a practical examination: Make a complete qualitative analysis of each of the substances A & B, and a quantitative analysis of one of them. Time allowed—five days, six hours each day.

K. M. Reese
Chemical & Engineering News
Vol. 60, p. 56, Aug. 23, 1982

Chemistry Fit for Ladies

Chemical News & Journal of Physical Science was a weekly founded in England in 1859 by Sir William Crookes and edited by him until 1906. This journal carried a weekly feature "Answers to Correspondents." In this column for July 14, 1865 appeared the following response:

Lucy asks if there have been any female chemists. Miss Chenevix was, we believe, the only lady who ever exercised her abilities on the products of the test tube and the crucible. There is no reason why ladies should not study and practice chemistry, and there are many reasons why they should choose this science to occupy their leisure moments. Surely the investigation of chemical facts would be more suitable for ladies then the study of social science, surgery, and medicine.

> **K. M. Reese**
> *Chemical & Engineering News*
> Vol. 60, p. 40, Jan. 4, 1982

Elementary Chemistry of Long Ago

> *The Young Ladies's and Gentlemen's Seminary*
> King St., New York

Messrs Dayton and Newman:

After a careful examination of your "Metrical Stories in Chemistry and Natural Philosopy," I do not hesitate to give it my unqualified approbation. It is a fact all must acknowledge that our youth are lamentably deficient in chemical knowledge. Frequently we see boys and even girls studying Botany and Astronomy, to the entire neglect of Chemistry. Now this should not be. A knowledge of chemistry is far more important for children that expect to have any concern with domestic affairs than Botany, Astronomy, or Mineralogy. The truths contained in your Book, and its cheapness, are recommendations sufficient to place it in the hands of every youth....

P. B. Heroy
New York, May 25, 1842

Such was the recommendation of the little volume *Metrical Stories in Chemistry and Natural Philosophy*, designed for youth by a teacher. It contains 144 pages of doggerel of various grades of quality. A few titles of these verses will give an idea of the general scheme of instruction.

First we have a rhyme about "Chisistry"; then "Matter"; "Attraction"; "Affinity"; several verses about "Light"; a number about "Caloric"; interspersed with tales, then a number of verses about electricity, gases, and various simple inorganic substances. Many of the tales contain the inevitable moral that used to be considered necessary in any story for youth. Questions are also asked at the end of each poem dealing with the subject matter. In this respect the book resembles modern texts....

How simple was the old theory of indicators.

> Within a glass of water put
>   A fair blue violet,
> Blue cabbage leaves, or radish root,
>   A clear, bright blue you get,
> Then pour Sulfuric Acid in,
>   And all the blue is fled,
> You'll see the change will quick begin,
>   Till all is turned to red.
>
> The acid is like water—white—
>   Why did it change the hue,
> And make the red appear so bright,
>   That was so fair a blue?
>
> Because they mix another way,
>   When we the acid pour,
> And they reflect a different ray
>   From what they did before.

Thermochemistry likewise presented few difficulties:

Caloric

Caloric is a subtil fluid that exists in every thing. It causes the sensation of heat, just as sugar causes a peculiar sensation, called sweetness. It cannot be seen, and is known to exist only from its effects.

Caloric—The Cause of Heat and Cold

Caloric is the element of heat. It goes from a hot body to a colder, till they are both of the same temperature. Its natural tendency is to make all things equally hot or cold.

Domestic economy is also taught:

> Molasses and wine
>   Expand much the more
> When Summer suns shine
>   Than when Winter winds roar.
> When Summer time passes,
>   And Winter comes round,
> Buy oil and molasses
>   They'll cheaper be found.

Beneath a woodcut showing "Helen's mother telling her of what water is composed, and how the little fishes live in it," is the following:

> Said Oxygen to Hydrogen,
>   "Let's meet—I think we ought to,"—
> They scarcely had united, when
>   They strangely turned to water,
>
> Though Hydrogen is twice as big,
>   When joining with the other,
> Yet Oxygen don't care a fig--
>   Outweighing eight times over.

The little fishes in the flood,
    That sport around so gayly,
Suck Oxygen to make their blood—
    And they must have it daily.

When iron gets wet, as oft it must,
    The Oxygen will meet it,
And cover it with a coat of rust,
    Because it likes to eat it.

And since the two have joined their hands,
    Their choicest gifts to scatter,
The earth with all her seas and lands
    Has been supplied with water.

As a last sample, the properties of sulfuric acid:

When Sulfur takes its greatest share
    Of Oxygen within it,
Sulfuric acid starts up there,
    Perfected in a minute.

Of all the acids on the globe
    This is the greatest glutton,
The slightest drop will eat a robe,
    And gnaw a metal button.

'Twill bite an iron bar in two,
    As people do a radish,
And one small drop on me and you,
    Would make us rather sadish.

If on your skin you get a drop,
    You should not pour on water,
It will not make the burning stop,
    But makes it rage the hotter.

But put some potash on the spot,
    Or chalk or even ashes,

'Twill gradual die till not a jot
      Is left to send its flashes.

If I should ever swallow it,
      (I'd not for mines of specie)
The safest remedy I'd get,
      Is chalk or else magnesia.

W. V. Sessions
Journal of Chemical Education
Vol. 3, 1288, 1926

Modern Physics Exam in Limerick Form

An exam in modern physics that a colleague and I cast in limerick form—the student must fill in the couplet or else the last line—has become part of the underground literature in physics instruction.   Countless editors, it seems, backed off from publishing it at the last minute. The exam is about physics; it is even literary.   But, apparently, in the eyes of the reader, it is too suggestive to take seriously. Her are a few test items:

An atom that came from the tap
Had electrons all over her map
      But in her interstices
      Lurked a much worse disease

- - - - - - - -

(That refers to radioactive decay, of course

Said a slow little neutron ere fission
"Don't speak of me with such derision"
      "I may have no charge
      and not be so large

- - - - - - - -

(That refers to nuclear forces, of course)

When a smart little wave named Swoom
Found a particle up in her room
    She remembered De Broglie
    And the Scriptures so holy

- - - - - - - -

(*That refers to the wave-particle quality, of course*)

An electron that was quite debonair
Spied a positron up on the stair

- - - - - - - -

- - - - - - - -
And finished him off in midair.

(*That refers to pair annihilation, of course*)

Perhaps I should tell you the sort of legitimate answer I
had in mind. On this one for example, remembering that the
electron and positron circle about their common center of mass,
forming a short-lived positronium atom before mutual annihilation, the student might have used a couplet like

    Starting the death dance
    She put him in a trance

Or, today, an incorrect answer, for which I'd give part
credit, might be

    She meant him no harm
    But turned on her charm

        . . . . and so it goes.

Robert Resnick
Oersted Medalist Lecture
AAPT-APS Meeting, Anaheim, December, 1974

Boners, Misnomers, and Bright Sayings

Among Astronomy Students

- A degree of latitude is shorter than it is at the pole.
- Polaris is the name of a star in the constellation of Asia Minor.
- A comet always keeps its tail as far away from the sun as possible.
- Civil time is that time adopted by all cilvilized countries.
- Name one of the requirements for the fulfillment of a total eclipse. *Answer:* A clear sky.
- Novae, or new stars, are stars that are here today and possibly gone tomorrow.
- A flash spectrum is caused by a meteor.
- Reference ws made to: "The nebulous (nebular) hypothesis and the planetarium [planetesimal] theory."
- Plato was the most recently discovered planet.
- As a comet nears the sun it undergoes spontaneous combustion.
- What does the name "Perseids" suggest? *Answer:* Andromeda's boyfriend.
- This course in astronomy has given me a hobby as I really enjoy the consolations at night when I'm not with a girl.
- Astronomy has been an interesting and apprehensive course.
- Pedigree is the point on the moon's orbit closest to the earth.
- Science is material. Religion is immaterial.
- What happened when the falling apple hit Newton on the head? *Answer:* He realized the gravity of the situation.
- Definition of infinity: "Out of this world."
- The comets' paths are hydrobolic, attabolic, and epileptic.
- Name an equatorial constellation. *Answer*: O'Ryan.
- What is the astronomical name for the year adopted in our present calandar reckoning? *Answer:* Year of confusion.
- The first signs of old age in a star are the swelling and reddening of its outer parts.
- The ancients placed mermaids, she-goats, virgins, and other legendary creatures in the zodiac.

Contributed by **Louis Berman**

Chemistry

- Hard water is water in which it is hard to dissolve things.
- A base is that substance that can stand by itself, that is, it has stability.
- Air cannot be a chemical compound because chemistry was not invented until thousands of years after the creation.
- Nonmetals may not be hammered.
- Sulfur is heated in Sicily and pumped out in Louisiana
- Supersaturated solution: one that contains more than it can hold.
- Nitrous oxide is called laughing gas because it is used by dentists.
- Water glass is the glass used for making tumblers.
- Galvanized iron is the iron used for making galvanometers.
- Gunpowder is mixed thoroughly so as to fool the ingredients into thinking they are a compound.
- An example of "reduction" is when water is boiled down it is reduced to a smaller amount after a bit.
- Temporary and permanent hardness of water signifies whether the water is in the form of snow or ice.
- Organic is the living germs in water and inorganic is the dead ones.
- When heat is applied to a gas, the molecules are stimulated to get away from the heat.
- A calorie is the amount of pressure required to push 1 gram of water 1°C.
- Molecules bump together and caused the molecular hypothesis.
- The phlogiston theory is a theory that used to be believed by chemists and people.
- A catalyst is a substance that felicitates a chemical change, but does not change itself.
- The late heat of vaporization is that which comes after the substance vaporized.
- Calcium chloride is a very thirsty substance.
- Sulfurated hydrogen is a gas of pugnacious odor.
- An ore is a mixture of various metals. A mine is an example.
- Hydrofluoric acid itches glass.

- Requisitions at the stock room window: alimony, porcelain
  vestibule, hot bath, pariscope, methyl orange juice,
  brunette with meniscus, iron fillings, hood, set of
  atomic weights.
- Gases transfer heat usually by conviction.
- A formula is a group of letters that can be broken up into
  elements.
- Iron occurs chiefly in Minnesota and in the blood.

Journal of Chemical Education
Vol. 2, pp. 408, 1196, 1925 and Vol. 3, p. 103, 1926

As General Hutton Said...

To the jaded examiner toiling through a pile of geological examination scripts, a good howler seems an oasis. The paper grader's mind is refreshed, a flight of fancy may be indulged in, before the task is resumed.

- General Hutton said, "The present is the key to the past."
  The average person does not have to dig a deep hole to be
  reminded of the past.
- At one time Wales was a steaming jungle.
- The enormity of geologic time.
- The rock appears to disappear.
- If a suspected criminal claimed to be in one place when the
  crime had been committed in another, a heavy mineral
  study of the clay and on the suspect's boots would
  soon prove who was right.
  Unfortunately, most criminals would clean their shoes.
- We have come a long way since Wagner proposed a flight
  from the poles.
- The North Sea is salt because of the Yarmouth bloaters.
- The dating of rocks depends very much on the superstition
  principle.

- Sexontidized maps enable you to locate your position more accurately.
- When India collided with the Asiatic block, the Himalayas were formed; when Africa merged into Europe, the North and South Downs were formed.
- Kaolin is metamorphosed feldspar, very hard and resistant, frozen stiff during the ice age and never recovered since. It is found in moraines and burns the fingers because it is so hot.
- If you see a big boulder that looks as if it came out of the sky, the chances are that it is an erratic.
- Evidence of unconformity...In a map, the contour lines cross one another.
- According to the experts, computers and Jehovah's witnesses are about to have another ice age.
- A change in basic rock time.
- Frank confessions of ignorance are often rather appealing, as in the answer "The coloring matter of emerald has been forgotten, but the writer does not think it is manganese," while one sympathizes with an unfortunate overseas student who had obviously been cramming for dear life, and broke down in midstream, with the remark "I have learned too much too quickly, and can no more think of nothing at all." One student notion, however, that the singular of specimen must clearly be speciman deserves mentions.
- I would like to think that the English and writing in no way detracted from the understandability of the report, and that any terminology used was adequate relevant. Thank you for reading this report.
- Doldrums are a series of high rocks near the Equator.
- Hutton said that the present was the result of the past, which he called Uniformitarianism. Verner, on the other hand, preferred to believe in phenomena like Noah's Ark and said that granite was a sedimentary rock.
- Isostasy is a principle of equilibrium that is supposed to have been discovered in Archimedes' bath. All chemical elements are dissolved in seawater. The explanation is that rivers have been carrying dissolved miners into the sea for millions of years.

- The continents are a triple junction between aerial, surface, and marine zones.
- Hot spots occur near seduction zones.
- Mountains are formed in erogenous zones.
- Plates on the Earth's serface are driven by infernal processes. A breccia may be formed by desiccation (dry rot) or by frost action (cold rot) or by solution (wet rot) or by the action of the sun (hot rot). It may happen by faulting in any kind of rock (all rot).
- In 1620 Bacon noticed the fitness of the continents.
- The petrifying of stones in springs can also be seen in the household kettle.
- Party to leave bus station, bus No. 18. Alight at Sea Corner, Highcliffe, and proceed along shore. Tea at Barton-on-Sea. Small chisels advised.
- The earth would have taken a long time to cool if it had not been for the Ice Ages during the Precambrian.
- Many orogenies begin early in some other part of the world and only affect Britain an era or two later.
- Laplace said there was a mass of gas just standing around doing nothing.
- The first law of geology is the law of supposition.

W. D. Ian Rolfe, Ed.
Geological Howlers
Geological Society of Glasgow

Weigh It with Music

A truly startling experience is talking after filling your lungs with helium. The shrill tones of a *tenore castrato* exponentially descended to a manly baritone as the helium is purged from the lungs. The even more striking effect of hydrogen gas was mentioned as early as 1830 by Silliman. He adds the cautionary footnote:

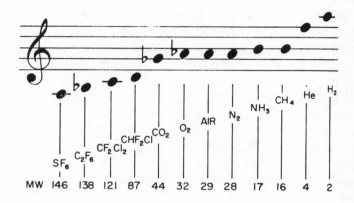

Figure 1. Relationship between pitch of note and molecular weight of sounding gas.

> Pilatre de Rozier was accustomed, not only to fill his lungs
> with hydrogen gas, but to set fire to it as it issued from his mouth,
> where it formed a very curious jet of flame. He also mixed pure
> hydrogen gas with one-ninth of common air, and respired the
> mixture as usual; "but when he attempted to set it on fire, the
> consequence was an explosion so dreadful, that he imagined his
> teeth were all blown out."

The words of Jacques in "As You Like It" are irresistibly
brought to mind:

> ...and his big manly voice,
> Turning again toward childish treble, pipes
> And whistles in his sound. Last scene of all,
> That ends this strange eventful history,
> Is second childishess and mere oblivion,
> Sans teeth, sans eyes, sans taste, sans everything.

Sulfur hexafluoride gives a profound deepening of the
voice; but even though it is used clinically for shadowing the
lungs in X-radiography, its inhalation, together with that of
hyrogen, is not recommended even by those chemists who
have total faith in the precedence of kinetic over thermodynamic
control of spontaneous reactions. Some even question the

breathing of helium and it is certainly unwise to feed xenon to goldfish. What is the cause of this frequency modulation from gas inhalation? The physics of the human voice is extremely complex, but there is little doubt that a major factor in the dramatic effect of helium is the much higher speed of sound in helium as compared with air.

We decided to test the generality of the effect. We filled large balloons with various gases and allowed them to sound through a small wooden organ pipe; a "B-flat D'Am" pipe served admirably. The open-ended pipe was first flushed, with a finger on the sounding slit. We then closed the end of the pipe and allowed the sound to stabilize. A musical ear can gage the pitch directly; tin ears should tape the sound and establish the pitch later using a standared pitch pipe. Our results are shown in Fig. 1. The distinctions are sharper (or flatter) than the coarsely quantized standard musical notation might suggest. Even the moderately tone-deaf can distinguish between air, nitrogen, and oxygen when the pure gases are sounded live against the recorded standard air pitch. Some ears are sensitive to 0.04 semitones, and would be able to detect 1% of oxygen in nitrogen or vice versa. The same principle has been applied as a whistle detector for gas chromatography, using ultrasonics in place of audible sounds, with the obvious advantages of miniaturization and piezocircuitry.

The data in Fig.1 clearly show a qualitative relationship between molecular weight and the pitch or frequency of the sound. Such frequencies may be used together with the velocity of sound in air (331 meter/sec) to derive values for the velocity of sound in the different gases. These values are in generally poor agreement with literature values obtained by more sophisticated means. The problem seems to be the back diffusion of air into the pipe. As early as 1829, Dulong attempted to minimize this by sounding an organ pipe inside a large chest filled with the gas. It is possible to sound a pipe inside a gasfilled glove bag, but the law of diminishing returns soon calls a halt. After all, there are far better methods available for the accurate determination of sound velocity in a gas.

Do such demonstrations have a pedagogic value? Indeed they do. How does the lecturer's voice reach the back row of the auditorium? Obviously, it is "carried" there by the molecules of the air; in a non-McLuhan sense, the medium is the message. If the molecules are carrying the message, clearly they cannot do so more rapidly than they are, on the average, moving. Newton was the first to attempt the calculation of the speed of sound in air. The value he obtained using the isothermal (Boyle's law) bulk modulus was substantially lower than the crude experimental value being used. His attempt to adjust the descrepancy is described in an article titled "Newton and the Fudge Factor" and subtitled "Fiddling with Sound." Much later, Laplace pointed out that the rapidity of sound-wave compressions and rarefactions necessitated the use of the adiabatic rather than the isothermal bulk modulus. A rather simple derivation leads to the following formula for the velocity of sound, U, of a gas of molecular weight, M:

$$U = [(\gamma RT)/M]^{1/2}$$

This is to be compared with the more familiar expression for the root mean square velocity of molecules of the same gas:

$$(\overline{u^2})^{1/2} = [(3RT)/M]^{1/2}$$

Since γ varies from about 1.15 to 1.66 for common gases, a knowledge of U and the assumption of a value of 1.4 for γ enables the molecular weight of an unknown gas to be estimated at about $\pm 10\%$.

The effect of hydrogen or helium upon the speaking voice is a fine example of the versatility of certain lecture demonstrations. It may be done merely for effect, or for laughter. And why not? Cakes and ale have their place, even in chemistry. But with a little time and effort, it may also be coaxed into yielding much insight into the interconnectedness of natural phenomena. Shakespeare's Jaques again makes the distinction between the two approaches best:

My lungs began to crow like chanticleer,
That fools should be so deep-contemplative,
And I did laugh sans intermission
An hour by his dial.

To which he counters:

> And then he drew a dial from his poke
> And looking on it with lack-lustre eye,
> Says, very wisely, "It is ten o'clock;
> Thus may we see," quoth he, "how the world wags."

"Seeing how the world wags." There could be worse definitions for the scientific enterprise.

D. A. Davenport,
M. Howe-Grant, and V. Srinivasan
Journal of Chemical Education **56**, 523 (1979)

How I Learned to Stop Worrying and Love Lecturing

Building a Professional Reputation can be done in two ways: You can create a technical or scientific achievement so outstanding that men will bring laurels to your door. Or else you can go out and actively seek professional distinction before you have anything tangible to your credit; in other words, make a career out of building a Professional Reputation.

Being a realist, I recognized early in my career that it was unlikely that I would ever make an outstanding technical contribution. Since I coveted the material rewards and personal satisfaction that come along with a Professional Reputation, I realized that my energies would best be devoted to the building of my professional standing.

Looking back on what I consider to be a modestly successful career, I attribute whatever success I enjoyed to my early realization that it was not *what* you said that was important but rather *how* you said it.

I had occasion at the very beginning to deliver a paper on a modest bit of market research I had done. While reading this paper, I had occasion to interject my personal misgivings about the limitations of market research and the problems of making our conclusions seem plausible to a skeptical top management.

After the paper, several colleagues came up to congratulate me. "You have put into words," they said, "what many of us felt but were afraid to say." I obviously had touched a responsive chord. Whereas my colleagues had assumed themselves unique in their doubts and feelings of inadequacy, I had assumed that my personal experience was representative and had generalized accordingly. This projection of personal experience, I realized, could make me a "spokesman for the profession."

So I began to devote my attention to methods for projecting my personality to an audience. I soon dropped the soporific habit of reading my papers; the speaker who reads clings to his manuscript rather than his audience.

Listening to other speakers, I noted that only by forming negative comments could I follow a paper that was read. Otherwise attention evaporates, the mind wanders. Since the reader pays no attention to the audience, the audience pays no attention to him.

"If I spoke off the cuff, I could not put it as concisely," the speaker apologizes to his audience. But the reason is his own anxiety, not concern for brevity. Without his manuscript he would get rattled, and the structure of his paper, often unclear to start with, might disappear altogether.

To avoid this pitfall, I took pains to create the impression of spontaneity. Though I always had a paper, and took it with me to the rostrum, I used it not as a crutch for myself but to avoid the appearance of intentional extemporaneousness. I would put the manuscript on the lectern and actually read a few lines, perhaps the introductory paragraph. But while doing so, I would look up more and more frequently, as if bothered by the way in which the manuscript seemed to get between me and my audience.

Soon I would begin to ingnore the text altogether, addressing myself directly to the audience. In this way, I achieved the convincing effect of spontaneity. Few people realized how practiced the performance was.

My output of papers grew tremendously as soon as I freed myself from the compulsion to believe what I wrote; I could then turn them out much more rapidly than my colleagues. This new freedom made my papers all the more successful because once I was relieved of the pressure to make the content significant, I could concentrate on making the manner intriguing.

As my reputation as a speaker grew, I received numerous requests to appear at meetings. I found that one paper was usually good for six or seven presentations. I could give it first at a local AIChE section, then at the local ACS. Usually there were a couple of other local groups that would ask to hear the paper, especially after I wrote to them indicating my availability. Finally I would present the paper at a national meeting; it would be published; and soon I would be freely dispensing reprints.

The screening procedures governing presentation of technical papers usually require that their content have some importance to the profession. But as one's reputation grows, whatever one chooses to say is assumed to have that vital importance because one does, after all, have a good reputation.

Of course, he who would pursue such a course cannot be universally liked. Frequently I would hear derogatory comments about me; I was called an exhibitionist, even a charlatan. Had the content of my papers been important to me, I would have been hurt. As it was, I remained detached. I regarded ill will as inevitable and simply made it part of my work to neutralize it.

I always acted interested when told of the criticisms. Never would I make a counter-criticism, but rather would always seek a way to say something complimentary. "That sounds just like Jack," I would say. "He gets so emotional at times and, who knows, maybe I deserved it. But did you hear that paper he presented on cost analysis at New Orleans last year? Superb!" (I knew my comment would reach him.)

Then I would seek out my detractor and force him into a pleasant conversation. I knew that my attention to his ego would make him feel less free to make derogatory comments once we had parted.

In due course, I became a consultant and my services have always been in great demand.

The profession has been very good to me.

Ernest Grand
Chemical Engineering
p. 153, January 20, 1964

The Dr. Fox Effect

The lecturer was Dr. Myron L. Fox, an authority on the application of mathematics to human behavior.

His topic: Mathematical Game Theory as Applied to Physical Education.

His credentials: Impressive.

His audience: 55 medical educators, psychologists, psychiatrists, and educational administrators.

The only problem with the above scene, which actually took place, was that Dr. Myron L. Fox was a fraud—a professional actor decked out with phony degrees and publications to seem respectable.

He had been coached to present his topic and conduct the question-and-answer period "with an excessive use of double-talk, neologism and nonsequiturs, and contradictory statements."

"All this was to be interspersed with parenthetical humor and meaningless references to unrelated topics," reported his coaches.

Dr. Fox was part of an elaborate scheme devised by three medical educators to find out whether the audience could be seduced by the style of the presentation.

They were.

Not one of the 55 victims of the hoax recognized it. One of them thought he had read Dr. Fox's publications.

Even so, not all of the victims were impressed with Dr. Fox. One thought the presentation was "too intellectual"; another described him as being "somewhat disorganized."

But overall, reported the authors of the study, D. H. Naftulin, J. E. Ware, Jr., and F. A. Donnelly, the 55 subjects "responded favorably at a significant level of an eight-item questionnaire concerning their attitudes toward the lecture."

Their hypothesis was that given a sufficiently "impressive lecturer and environment for the lecture,"even experienced educators participating in a new learning experience can be seduced into feeling satisfied that they have learned, despite irrelevant, conflicting, and meaningless content conveyed by the lecturer."

Journal of Medical Education
Vol. 50, February 1975

Student Ratings

Colleges and universities are employing student evaluations of faculty with increasing frequency. The optional use of student ratings in order to improve instruction often evolves to the mandatory consideration of student rating data in making decisions about faculty rentention, promotion, and tenure.

In establishing the validity of a student-faculty evaluation questionnaire, one must answer two questions. First, is the evaluation instrument sensitive to differences in instruction? Second, does the instrument provide a rating that is a valid index of overall instructional effectiveness?

Subjects in the "Dr. Fox" study evaluated Dr. Fox's lecture favorably on an eight-item satisfaction questionnaire. These findings were taken as evidence that students and faculty may evaluate lectures favorably even in the absence of sustance in lecture presentations. This phenomenon was characterized as an "illusion" of having learned. However, this conclusion does not follow directly from the data given because listeners were not asked to rate learning gain, and no measure of achievement was employed.

The present study was designed to provide an experimental test of the effects of lecturer seduction and content coverage of student ratings of instruction and student achievement....

The results of the present study suggest that the ratings of students exposed to low seduction lectures more accurately reflect differences in content coverage than do ratings of students exposed to high seduction lectures. Under high-seduction lecture conditions, student satisfaction ratings were generally not sensitive to either differences in content coverage or actual test performance.

The use of student ratings to make decisions regarding faculty retention, tenure, and promotion may be invalid. Faculty who master the "Doctor Fox Effect" may receive favorable ratings regardless of how well they know their subjects and regardless of how much their students learn.

The phenomenon of educational seduction appears to be more complicated than was originally thought. Naftulin, Ware, and Donnelly were correct in anticipating that teaching effectiveness may be optimized with high levels of content coverage *and* a seductive presentation manner. On the other hand, the Doctor Fox Effect appears to be much more than an illusion. Whereas teaching style is a major factor in determining student ratings, it is also a powerful influence of student test perform-ance.

J. E. Ware and **R. G. Williams**
Journal of Medical Education
Vol. 50, pp. 149–155, February, 1975

American Geophysical Union
Member Self-Evaluation Test

As a service to the membership of the AGU, the Self-Improvement Committee (Sic) has designed the following self-evaluation test to help you determine where you are in your career. Please do not be afraid to take it. You cannot fail this test even if your career is a failure. Come now, pull yourself together, pick up a pencil, and circle the number corresponding to the statement that best describes you. Please be honest; we have ways of detecting cheaters. Do not read through the test once first; you must answer the questions the first time you read them. There shall be no deviations from these rules!

When the test is finished you may clip the test from the journal. Then throw the test away and save the journal. It would make Fred Spilhaus, our editor, very happy to know that one issue of *EOS* had been saved!

Section A. Employment

I. 1. Have no office.
    2. Have office, no phone.
    3. Have office with phone.
    4. Have office and a secretary to answer the phone.
    5. Receptionist answers phone for secretary.
    6. After receptionist screens callers, secretary will take call if not too busy.

II. 1. Do not receive phone calls.
     2. Salesmen occasionally call.
     3. Two or three calls a day.
     4. Half day on phone.
     5. Phone constantly busy; cannot be reached.
     6. People have given up phoning.
III. 1. Leave self messages.
     2. Return all calls.
     3. Return calls from people I know.
     4. Return calls from funding agencies.
     5. Return calls from White House
     6. Make calls from White House

Section B. Reading Habits

IV. 1. Read all articles in JGR.
     2. Read all abstracts in JGR, then articles of interest.
     3. Read all titles, then abstracts of interest.
     4. Scan author list; read own papers.
     5. Unwrap journal; file it.
     6. Do not unwrap journals nor even unfold *EOS*.
V.   1. Read abstract, then body of paper.
     2. Read abstract; scan article; then read conclusions.
     3. Read authors, then conclusions.
     4. Read authors, then references.
     5. Read references; guess at authors.
     6. Read references only; count citations.

Section C. AGU Association

VI. 1. Student member.
     2. Full member.
     3. Serve on AGU committees.
     4. Elected to AGU office.
     5. Known too well to be elected to AGU office.
     6. AGU Fellow.
VII. 1. Author in non-AGU journal.
     2. Author in GRL.

3.   Author in JRG.
4.   Author in RGSP.
5.   Associate editor of AGU journal.
6.   Editor: publishes in non-AGU journal.

Section D. Publication

VIII. 1.   Do not publish yet.
     2.   Principal author all papers.
     3.   Coauthor some papers.
     4.   Write a few papers, mainly coauthor.
     5.   Only coauthor.
     6.   Do not publish anymore.

IX. 1.   Never asked to review papers.
    2.   Occasionally asked to review papers.
    3.   Often asked to review.
    4.   All too often asked to review.
    5.   Reject any and all papers.
    6.   Never asked to review.

Section E. AGU Meetings

X. 1.   Occasionally attend AGU meetings.
    2.   Usually attend one of Spring and Fall meetings.
    3.   Attend both Spring and Fall meetings.
    4.   Attend Spring and Fall meeting and all Chapman conferences in field.
    5.   Attend all of above and an international conference or two.
    6.   Fly from meeting to meeting throughout year.
XI. 1.   Find interesting session and sit through it all.
    2.   Carefully choose sessions, may switch at coffee break.
    3.   Run from session to session catching. interesting talks.
    4.   Stop to talk to colleagues between interesting talks
    5.   Register for meetings; stand in hall.
    6.   Do not attend meetings.
XII. 1.   Find all talks interesting but understand few.

2. Find something interesting in most talks; under-
    stand more.
3. Find a few talks interesting; understand most talks.
4. Find own talks interesting
5. Lose interest in even own talks.
6. If authors have had time to prepare a talk, results
    are too old to be bothered with.

Section F. Proposal Writing

XIII. 1. No one knows me.
2. Write proposal; someone else is PI.
3. Am PI; someone else wrote proposal.
4. On committee to define mission.
5. Chair committee to judge proposals for mission
    I defined.
6. Chair committee that writes report saying my
    favorite subject is the most outstanding problem in
    the field for the next 10 years.

XIV. 1. Have never written a proposal.
2. Named in someone else's proposal.
3. After months of negoiation with department chair-
    man and dean, send off proposal to NSF
    for $10K.
4. Proposal to NSF for $25K sits on dean's desk for
    less than a month; only minor comments.
5. Multiyear, $100,000-dollar proposal to NASA;
    dean phones to give his best wishes.
6. Million dollar proposal; taken out to dinner by
    University president. Dean reads about
    proposal in newspaper.

Scoring: Method A. Give yourself one point for each level 1
response; two for level 2; three for level 3; four for level 4, five
for level 5, etc. Add up to get total number of points, See key
below. Method B. Convert responses to machine readable format; write on 1600 bpi, 9-track, BCD format; enclose $50;
send to author. *Key:* 0–13 points: You cheated! The minimum score is 14 points. Go to scoring method B. 14 points.
You are obviously a student member without an office or a

phone. Be thankful you still have time to read all the JGR. It won't last long. 15–17 points: Your career is beginning to develop. You are on your way. 28–41 points: No stopping you now. 42–55 points Midcareer: Now its downhill! 56–69 points: You can relax now and look back over a productive career. 70–83 points: The ultimate level. Once a year go to the library and count your citations in the *Science Citation Index*. 84 points: You cheated! How could you be filling out this test and have answered question IV-6 affirmatively. Follow modified Method B—send blank tape and $50 to author.

Christopher T. Russell
EOS, 60, No. 49, p. 1024
December, 1979

A Nonsense Book of Thermodynamics

There was a young man named Carnot
Who fervently wanted to know
How caloric dispersed
In cycles reversed
Made engines efficiently go.

Helmholtz, Mayer, and Joule
All found the very same rule:
That the preposterous notion
Of perpetual motion
Is the pipe dream of only the fool.

Clausius found Law Number Two:
"Die Entropie strebt einem Maximum zu."
"Energy dissipate,"
 Cried Peter G. Tait.
"Yes," said Lord Kelvin, "Pray do."

Take (said Maxwell) a common foot rule
Like that you had while at school.
By demoniacally sorting
The molecular cavorting
One end is distinctly more cool.

Boltzmann (who was one of the best)
Applied a statistically significant test.
To equilibrate
Is everyone's fate
And to H with all of the rest.

Gibbs found the famous phase law
Which works with nary a flaw.
"F equals c
Plus 2 minus p."
So reads the famous phase law.

Nernst found Law Number Three,
Which he expounded with infinite glee,
"As zero you near
Let dS disappear
And dH will equal dG."

Contributed by **William B. Jensen**

Tellurium Via Dysprosium

It is often a difficult thing to say no. To do so grace-fully requires a great deal of wit, wisdom, and ingenuity. Below is a small classic of how one former Harvard student (we presume in chemistry) gently declined contributing to the Harvard Fund at the close of the depression.

Halogens,

As I am an aluminum of two colleges besides Harvard, and cannot, with my bismuth in its present state, pay antimony to all three, I hope you will not think me a cadmium if I do not cesium this opportunity of making a donation. So far this year I have metal current expenses, but in these troubled times when the future holds in store we know not phosphorus, I could not make a contribution without boron from the bank. It would nickel out of my savings. A manganese spend the dollars these days; a tin spot is gone in no time. You're lead to feel you're pouring them down your zinc; we, arsenic. Much better to sodium up in a stocking. So don't be silicon not make any contribution this year unless a bromine helps me out.

Very unruly yours,

G. Alex Mills
University of Delaware
Humor Among Chemists, 1962

The Clerihew

We have just learned, to our dismay, of a form of light verse called the Clerihew, after its inventor, Edward Clerihew Bentley. Here is the first Clerihew committed (in the early 1900s) by Mr. Bentley.

Sir Humphry Davy
Detested gravy.
He lived in the odium
Of having discovered sodium.

More on the Clerihew

Dear Sir:

So! *Chem and Engineering News*
Has just now heard of Clerihew
I trust that *JACS*
was not so lax.

* * *

Phase Rule

A scientist, named Tweedle Deed'm,
Refused to let phase rule impede'm
So he married a wife:
The new phase in his life
Cost him a degree of his freed'm

* * *

- Avid chemistry student voted "most likely to dissolve."
- Whistler (who attended West Point): "If silicon had been a gas, I might have been a general."
- Q: What do chemists call a substance that helps a reaction, yet does not take part in the reaction?
  A: A capitalist.

G. Alex Mills

The Poker Laws of Thermodynamics

First: You can't win.
Second: You can't break even.
Third: You can't quit.

- Iron was discovered because someone smelt it.
- Sea water has the formula CH_2O.
- Glycerine is a vicious liquid, miserable in water in all proportions.

- A hydrate is a substance that contains water. Example: a watermelon.
- Q. Where is the world's largest helium plant located?
  A: I'm dumb, but at least I know it doesn't grow in a plant.
- Q: When you spill an acid on your clothes, what do you see?
  A: Your skin.
- A molar solution is one that has been thoroughly chewed.
- Quinine is the bark of a tree; canine is the bark of a dog.
- An axiom is a thing that is so visible that it is not necessary to see it.
- The metric system refers to kilograms, centigrams, telegrams, etc.
- The big toe is sometimes called the pedagogue.

G. Alex Mills

Little Ion

"Little Ion in my flask
Do you mind much if I ask
What your name is, Little Ion,
Can't you see you've got me cryin'?
"Can't you see I'm growing weaker
As you hide there in my beaker
Ain't you got no heart at all?
Don't you care if I flunk Qual?
You could stop my endless trying'
To find your name out, Little Ion.
You could end all my confusion.
If you'd come out of solution.

Jerry Wellins

Traitmarks

How you can tell someone's a chemist:

1. A chemist's old pipe is burned away on one side. more than the other from lighting it with a Bunsen burner.
2. The chemist, stirring his morning coffee, always touches the tip of the spoon to the inside rim of the cup.
3. A chemist holds a highball glass at eye-level when putting in the liquor instead of using a jigger.
4. A chemist pronounces iodine so that it rhymes with bean, and salicylate with accent on the first syllable.
5. The chemist's house abounds with Erlenmeyer and Florence flasks used as vases, decanters, and vinegar cruets, and has evaporating dishes for ash trays.

Linus M. Web

Collected by **G. Alex Mills**
University of Delaware
in *Humor Among Chemists*
Privately printed, 1962

Filling in the Spaces

The following example of the art of programmed instruction was prepared by Mr. Lackey of the US Naval Dental School in Bethesda, a professional programmer, for the edification of his colleagues. It originally appeared as a "workbook" consisting of what programmers call "frames"—a series of cards arranged in sequence.

(1)

HELLO THERE!

We want to show you a sample of your new lesson guide. You will be using this new lesson guide and we think you will be happier if you know how it works. So please go on to the next card.

(2)

VERY GOOD!

We asked you to go on to the next card, which is this one, and you did.
That's just fine.
You are doing very well.
Now please go on to the next card.

(3)

FINE!

Here you are on the third card already.
Now we can start the game. Here is how we play it.
Each card has lots of words on it, telling you something. But some of the words will be left out, and you will have to fill them in. Like on the next card.
Go on to it, please.

(4)

DANDY!

So when you see a space where a word should be, you fill it in
With your pencil.
When   the   word   is   left   out,   you   will   see   a   space
_____
Write the word in the_____
And go on to the next card.

(5)

WHOOPS!

Did you write anything in the _____?

You were bad and did not follow instructions.
You must not think you are so smart.
You must do what we say and write the w_____
in the space. And then go on to the next card where you will
see the missing word at the top.

(6)

WORD

That is very good.
Did you write it tiny so that it fit the space?
All right. Now we will sum up what we have said and get on
with the lesson guide.
Next card pl_____

(7)

PLEASE

On each card of the lesson guide, there will be a missing word.
Where the word is missing, there will be a space for you to fill
in.
You will go from card to card, filling in the spaces. With
words.
Perhaps you've caught on be now. You f_____ in the
spaces.

(8)

FILLED

Perhaps, too, you are rather tired of filling in the spaces.
Perhaps you think you don't learn so much by f_____
in the sp_____

(9)

FILLING IN THE SPACES

Perhaps you thing there's something more to learning than just
filling in the spaces.
If so, you are a real s_____ o_____ b_____

(10)

SHREWD OBSERVER, BUSTER

Because there is.

Melvin W. Lackey
The Saturday Review
Sampler of Wit and Wisdom
Simon and Schuster, pp. 19–21, 1966

* * *

This man must be very ignorant, for he answers every
question he is asked.

Voltaire

Doodles from Scientists

The Prospect for Fusion

Professor Rose made this drawing in 1958; it was meant to adorn the program for the annual Gaseous Electronics Conference of the American Physical Society, but Melvin Gottlieb, director of the Princeton Plasma Physics Laboratory, feared that the federal government would take umbrage at too much humor. After 18 years, Rose included the cartoon in his paper in the MIT publication.

In 1958, research on controlled fusion had just been de-classified. The public was told that its advent was imminent. Yet even then, many problems were recognized. "E/P" repre-sents the ratio of electric field to pressure; it is a parameter crucial to the creation of a plasma—a gas of charged particles. "Instability" is short for plasma instability; in 1958 one of the most dreaded varieties was the so-called sausage instability, in which a plasma column pinches off at intervals, and bulges in-between. "Propagation" is that of electromagnetic waves; they bear upon the successful confinement of a plasma. The draw-ing's caption refers to a selling point (then and now) for con-trolled fusion: the prospect that a fusion reactor will be fueled by a mixture of deuterium (D) and tritium (T), both isotopes of hydrogen. The former, at least, is in essentially endless sup-ply, for it is a component of ordinary water.

David J. Rose
Technology Review
Vol. 79, 2, p. 20, 1976

Avoirdupois

The length of this line indicates the ton of coal as dug by the miner.
This one indicates the ton shipped to the dealer,
The smaller dealer gets a ton like this.
This is the ton you pay for.
This is what you get.
The residue is:
Cinders and Ashes.
And this line will give you some conception of the size of the bill.

Carolyn Wells
A Whimsical Anthology
Charles Scribner & Sons, NY, 1906

Astronomical Poetry and Quips

The Nebular Hypothesis

The stars are as thick as flowers in the sky;
Tonight is a night for lovers to kiss;
But we are arguing, you and I,
On the nebular hypothesis.

Mary Burwell

* * *

Hubble Bubble

When Jean grows too didactic
Or Friedman makes too free
Among extragalactic
Clusters and nebulae—
When life is full of trouble
And mostly froth and bubble
I turn to Dr. Hubble
He is the man for me.

Observatory
June 1931

* * *

Observatory

I saw a tutor take his tube
   The Comet's course to spy;
I heard a scream—the gathered rays
   Had stewed the tutor's eye!

Oliver Wendell Holmes
The Comet, 1832

* * *

What was God doing before He made Heaven and Earth?
He was creating Hell for people who ask questions like that.

St. Augustine

For a moment of night we have a glimpse of ourselves and of our world islanded in its stream of stars—pilgrims of mortality, voyaging between the horizons across the eternal seas of space and time.

Henry Beston
The Outermost House

There is nothing more incomprehensible than a wrangle among astronomers.

Henry L. Mencken

Ten years of radioastronomy have taught humanity more about the creation and organization of the universe than thousands of years of religion and philosophy.

P. C. W. Davies
Space and Time in the
Modern Universe

The astronomers said: "Give us matter, and a little motion, and we will construct the universe."

Ralph Waldo Emerson
Essay on Nature

Comets are the nearest thing to nothing that anything can still be called something.

Quoted in *National Geographic*
March 31, 1955

Of the real universe we know nothing, except that there exist as many versions of it as there are perceptive minds.

Gerald Bullitt
Dreaming

It is one thing for the human mind to extract from the phenomena of nature the laws which it has itself put into them; it may be a far harder thing to extract laws over which it has no control.

Arthur Eddington
Space, Time and Gravitation

Big whirls have little whirls
That feed on their velocity
And little whirls have lesser whirls
And so on to viscosity.

F. L. Richardson

Twinkle, twinkle little star
Out in space so very far,
If you're as bright as we think you are,
Beam a signal, little star.

Katherine O'Brien

Toy-maker Ptolemy
Made up a universe
Nine crystal yo-yos he
Spun on one string. It was
Something to see it go.
Half sad to see it pass.

John Ciardi
Some Sort of Game

I know perfectly well that at this moment the whole universe is listening to us—and that every word we say echoes to the remotest star.

Jean Giraudoux
The Madwoman of Chaillot

When I heard the learn'd astronomer,
When the proofs, the figures, were ranged in columns before
    me,
When I was shown the charts and diagrams, to add, divide,
    and measure them,

When I sitting heard the astronomer when he lectures with
     much applause in the lecture room,
How soon unaccountable I became tired and sick,
Till rising and gliding out I wander'd off by myself,
In the mystical moist night air, and from time to time,
Looked up in perfect silence at the stars.

Walt Whitman
When I Heard the Learn'd Astronomer

You know Orion always comes up sideways
Throwing a leg up over our fence of mountains.

Robert Frost
The Star-Splitter

. . . The learned astronomer
Analyzing the light of most remote star-swirls
Has found them—or a trick of distance eludes his prism—
All at incredible speeds fleeing outward from ours.
I thought, no doubt they are fleeing the contagion
Of consciousness that infects this corner of space.

Robinson Jeffers
Margrave

Analyzing a spectrum is exactly like doing a crossword
puzzle, but when you get through with it, you call the answer
research.

Henry Norris Russell

It is a funny thing that the Metropolitan Office can forecast
an eclipse 100 years hence, but cannot tell you whether to take
an umbrella with you the next morning.

Observatory
December 1935

Mrs. Frisch has kindly shared a cartoon by the late Dr. Otto R. Frisch. "The animal has obvious character and a somewhat hyperthyroid, startled expression, perhaps occasioned by the presence of its tail in the artist's sleeve," Charles McCutchen, one of Dr. Frisch's students, observed.

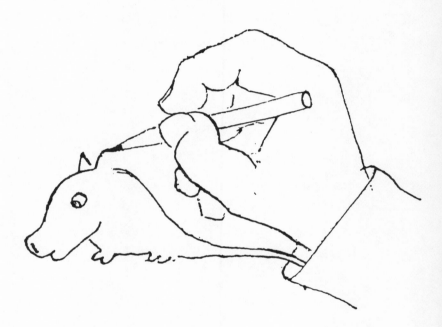

THE HIPPOPOTAMOUSE

Publish and Perish

The Definitive Article

I wrote the definitive article
And sent it to *Physics Today*.
It described every yet unknown particle
And explained every paradox away.
In elegant but simple prose
The writing flows and flows and flows
Everything with nothing missed
Of interest to a physicist.

The journal said it would not fit;
Let's undertake to shorten it.
Two hundred pages in precis
Is even long for RMP.
Fifteen pages—that will do
For the *Physical Review*.

But wait!
Boil it down, evaporate
Everything but name and date.
There. . . now we're set—
Send it off to *Phys. Rev. Lett.!*

Marilyn T. Kocher
Physics Today, October 1979

An Interview With A. A. Aardvark

Background. Recently, the Journal of Geophysical Research *(JGR) offices received a manuscript authored by an A. A. Aardvark. The Editor had never heard of this scientist but, since it was a reasonably well written article, assumed that a regular contributor had suddenly taken a nom-de-plume. Further, Aardvark's name could not be found in the paper's bibliography, a dead giveaway that Aardvark was not the writer's real name, for even anonymous authors reference themselves. Since the manuscript was postmarked from my city and since the author had requested that all correspondence be forwarded to a post office box there, it was suggested that I try to track down Aardvark and ascertain what had led to this action. No doubt the Editor just thought that it would be of interest to the American Geophysical Union (AGU) membership if I conducted an in-depth interview. Armed with a copy of the bibliography of Aardvark's paper, I found it quite easy to track the anonymous author. I knocked on Aardvark's door, identified myself, and waited. I could hear paper shuffling and file cabinets being closed; then the door opened a crack, and the interview began.*

AAA.   What do you want?
JGR.   The editor gave me a copy of your paper and asked me to talk to you.
AAA   Are you a referee?
JGR.   No. I just want to interview you about your name change.
AAA.   Are you a member of the AGU?
JGR.   Of course.
AAA.   What's the AGU motto?
JGR.   Unselfish cooperation in research. Why?
AAA.   Not many people know that, or at least remember. You're okay, kid. Come on in.

I was ushered into a very dimly lit room. The blinds were drawn, and, as I had expected, all the file cabinets were shut. Further, every bit of paper in the office was turned face down. I thought I would get down to brass tacks right away.

JGR.   Why the name change?
AAA.   Did you know I just published my one-hundreth paper?
JGR.   No, I didn't. Congratulations. (This was going to be a hard interview. He was avoiding the questions.)

AAA.   That's one hundred real publications. No abstracts, extended abstracts, unpublished reports, or any nonsense like that in there. One hundred real papers.

JGR.   That's really impressive. (I knew that meant he was prolific, but to a kid like me, 50, 100, 200—who could tell the difference?)

AAA.   One hundred real papers, and now they won't accept any more.

JGR.   You're kidding.

AAA.   No. I send in papers at the same rate as before, and they all get rejected.

JGR.   (Having received an occasional bad review myself, I was sympathetic.) They all get rejected? Surely some get through occasionally. When I come down hard on a paper, the other referee always says it is the best paper he had read in a decade. Occasionally, you must get two of the latter.

AAA.   No. Never!

JGR.   What has all this got to do with your name change? (I felt the interview slipping away from me, and I decided to get back to the main topic.)

AAA.   I have a theory that each controversial new idea that you publish will alienate some fraction of your colleagues. When you have published a critical number of papers, you finally have alienated everyone in the potential referee pool.

JGR.   But wait a minute! (I thought I had him. I'd taken freshman statistics.) Should not the probability of rejection gradually go to unity, so that you simply have fewer and fewer papers being accepted?

AAA.   That assumes that the refereeing process is random, but there is a Maxwell's demon involved, the Editor, who directs your paper to new neutral referees if he notices any antagonism appearing in the old neutral referees' reports. Eventually, he uses up the entire referee pool, and no more publications for you.

JGR.   Do you have a solution?

AAA.   Yes, change your name.

JGR.   (Thank goodness, he had returned to the subject at hand, I had just been about to ask again.) But why A. A. Aardvark? And what does the A. A. stand for?

AAA.   The second A stands for Aaron. The first A is simply
an initial. As for why Aardvark, the answer is
obvious.

JGR.   (I immediately saw the light and jumped to a conclusion, as I am wont to do.) You want to be first in
the reference list.

AAA.   No, that's not it. However, there may be some of that
going on. Let me show you something very interesting.

JGR.   (I was crushed. What could be more obvious and
natural than wanting to be first in the reference list?
He turned over one of the pieces of graph paper that
had been lying face down on his table. This graph is
reproduced with Aardvark's permission in Fig. 1).

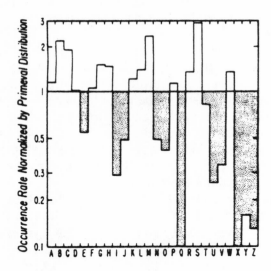

Fig. 1.   Illustration of the Aardvark distribution function. The
occurrence rate of the first letter of the last name is normalized by the
primeval distribution occurrence rate. Data were taken from a recent issue
of the *Journal of Geophysical Research.*

AAA.   This is the distribution of the first letter of authors' last
names in the mst recent JGR, normalized by the
primeval distribution, in which I assume that all names
were equally distributed throughout the alphabet.

Notice that in the first half of the alphabet, only E, I, and J do not have their fair share of last names. Just the opposite is true in the last half of the alphabet. Only P, R, S, and W have at least their fair share. In fact, P, R, and W just barely have their fair share. If we integrate this distribution function, we find that by the letter D, 30% of all references occur, whereas the primeval distribution would predict just half this number.

JGR. (Trying to recover from my early mistakes, I drew the obvious conclusion again.) You believe then that people tend to reference mainly authors in the top part of the alphabet, and by changing your name to begin with an A you will be referenced more.

AAA. I don't believe that, but I am afraid that many people do. I feel that for this reason the curve is a dynamic one and subject to a very elementary instability. It is dynamic because as people become aware of this distribution, and it is obvious to anyone who scans reference lists, there is a great impulse to change one's name to begin earlier in the alphabet. The really strong egos will opt for As, the lesser egos for Bs, etc. A recent example of this process is given by Hirshberg, who changed her name to Feynman. She is obviously a timid sort, since she advanced herself only 13%. The instability arises because as more and more people change their names, the distribution becomes more and more skewed, and the pressure to change one's name becomes greater. Eventually, everyone will have a last name beginning with A. However, I do not foresee that occurring in our lifetime.

JGR. And you're just trying to get a jump on the pack.

AAA. No, no. You missinterpret my motives entirely.

JGR. Do you really believe that people with names early in the alphabet get more favored treatment when it comes to referencing?

AAA. If there is an effect, it is probably small, but I can't prove it. Certainly people like Akasofu and Axford did not rise to the respected positions they now hold because their names started with A. However, it is not

what I believe that counts. My name begins with A's.
It is what the non-A's believe that leads to the instability.

JGR. (I felt he wasn't being quite honest with me. His
name hadn't always begun with A. I was beginning
to get irritated.) You are on an ego trip then. You just
want to get your name first in the reference list.

AAA. If I had been looking for reams of credit that didn't
belong to me, I would have changed my name to Et
Al.

JGR. (After that putdown, I decided to be a little more
prudent.). What about your promotions, tenure, etc.?
If you changed your name, you won't receive credit
for the work you have done will you?

AAA. Perhaps I'm being too cynical, but most of one's
immediate colleagues are quite unaware of what one is
doing. The people most aware of one's contrbutions
are those working on similar problems at other
institutions who have no influence in the promotion
procedure. When outside opinion is sought, those
colleagues who are unaware of what you are doing are
the ones who poll the outside world. What chance
have they of locating those who are really familiar
with your work?

JGR. (Since Aardvark appeared to be tiring of all this questioning and because the hour was late, I excused myself and headed for the door. But as I left, one last
thought occurred to me.) I admit that your theory
may work. The referees may start afresh, and your
professional career may be totally unaffected. But
what about when you have published a second hundred papers? What will you do then? May I talk to
you again, and how will I know when it happens?
(AAA's reply was very succinct.)

AAA. Just give me a call when you start seeing papers by
Barry B. Barracuda.

Christopher T. Russell
EOS 59, #3,
March 1978, p. 118

Epitaphium Chemicum

Here lieth to digest, maturate, and amalgamate,
    with clay,
In Balnec Arenae,
Stratum super stratum
The *residuum, terra damnata,* and *caput mortuum*
    of BOYLE GODFREY, Chymist;
A man who in this earthly laboratory
Pursued various processes to obtain
Arcanum vitae, of the secret to live;
Also *aurum vitae*
Or the art of getting rather than making gold.
Alchymist like—he saw
All his labor and projection,
As mercury in the fire, evaporated in fume.
When he dissolved to his first principles,
He departed as poor
As the last drops of an alembic;
For riches are not poured
On the adepts of this world.
Though fond of novelty, he carefully avoided
The fermentation, effervescense, and decrepitation,
Of this life.
Full seventy years his exalted essence
Was hermetically seal'd in its terrene matrass:
But the radical moisture being exhausted,
The elixir vitae spent,
And exsiccated to a cuticle,
He could not suspend longer in his vehicle,
But precipitated, *gradatim, per campanum,*
To his original dust.
May the light above
Brighter than Bolognian phosphorus,
Preserve him from
The athaner, empyreuma, and reverberating furnace
Of the other world!
Depurate him from the faces and scoria
Of this;
Highly rectify and volatilize

His aetherial spirit;
Bring it ever to the balm of the retort
of this globe;
Place it in a proper recipient,
Or chrystaline orb,
Among the elect of the flowers of Benjamin;
Never to be saturated
Till the general resuscitation,
Conflagration, calcination,
And sublimation of all things!

Anonymous
The Columbian Magazine
Vol. 3, 1789, p. 367;
The Journal of Chemical Education
Vol. 24, 435, 1947

The Author

Always keen to break new ground,
Our friend called Joe spent many days
On working out a new technique
For serum pussicatalase.
Now, he thought, there'll be, one day,
A paper out by Muggins, J.

Just what he'd done, with what result,
He wrote up in his neatest scrawl,
And waffled for a page or so
On why you measured it at all.
And then comparisons were needed
With methods he had superseded.

The next stage was a typescript draft—
The office said they'd do it soon,
On days when clinics weren't too bad—
A page each Tuesday afternoon.
And meanwhile, when he'd time to spare,
Joe drew his graphs with loving care.

With scissors, pen and, Sellotape,
He pulled his typescript draft about,
Read through the journal of his choice,
And did his best to set it out,
With headings all arranged, this time,
To suit the *J. Prolif. Enzym.*

This journal has its own odd style—
Although he'd tried to make this plain,
The typist hadn't understood,
So had to do it all again.
This time she'd done him really fine—
Just here and there left out a line.

The figures, taking many weeks,
The journal's rules precisely met
On size and shape and type of print,
With Standardgraph and Letraset.
Assembling all the bits with care,
He packed it off and said a prayer.

The editors were not impressed.
But since the method worked, said Joe,
Far better than the usual ones,
Well, other people ought to know!
And so it now became his duty
To find a journal not so snooty.

He chose one and rewrote his screed,
But soon was getting rather vexed,
The way they did their references,
As little numbers in the text.
He'd somewhere miss one out and then,
He'd have to do the lot again.

Initials first, then NAME IN CAPS,
Without a heading at the top,
<u>Journal underlined</u> in small,
The volume (year) first page full stop.
Accustomed to just typing prose,
This took the office several goes.

He packed it safely for the post
And heaved a sign of satisfaction,
Trotted back into the lab
And, starting up a fat extraction,
He saw his journal by the rack,
And glanced at what was on the back.

"Do not mount the figures, please."
He read it out and said: "Oh Lord!
The ones so neatly parcelled up
Are all shipshape on Bristol Board!"
Joe put aside his Soxhlet thimble
And made a rude two-fingered symbol.

"Unpack it all? Repeat the lot?'
He said: "No hemorrhagic fear!"
And sent it off by first-class post;
But what you people want to hear,
Is what the journal said to Joe—
"The editor will let you know."

Douglas E. Kidder
Verses Bright and Beautiful
University of Bristol, 1983

Researchers' Prayers

The Proceedings of the Chemical Society *records some of the
128 stanzas submitted in a competition at Christmas 1962 for
quatrains in the style of the Fisherman's Prayer: "God give me
strength to catch a fish/So large that even I/When telling of it
afterward/May never need to lie."*

O Lord, I fall upon my knees
And pray that all my syntheses
May no longer be inferior
To those conducted by bacteria.

LJB

Lord give me smartness to discern the work that's in the fashion,
Some quick results and facile thoughts for plausible discussion;
Then when I've rushed my paper in to claim priority,
Lord, spare me the attentions of a thorough referee.

"National Cynical Laboratory"

Let us hope he avoids RW (probably a case-hardened referee) who replies:

God grant there is never a paper
So good that even I
Without a stroke from my pencil
Will have to let it by.

RW

Oh, send us all the joy Divine,
The thrill of hope, so shy and fine:
Through oil and tar a tiny sign
Of cosmic order—crystalline.

(erg)

Oh Lord, when I lay down my load
And terminate my stint,
May I find peace in my abode
And be the first to print.

(From Belgium)

From Proceedings of the Chemical Society
January 1963, pp. 8–10

My Brother, the Author

It is not usual policy for this column to review books and this is only done when the book is written by brothers, wives, or intimate associates of the columnist.

Today's book is an impressive volume that is must reading for all guests at Haverford State Hospital. Written by John M. McCormick and Mario G. Salvadori, Numerical Methods in Fortran *(Prentice-Hall, 324 pages) is the first book to deal courageously with Fortran, a problem unknown in our culture 100 years ago.*

Mr. McCormick is an adjunct associate professor at Columbia University. Mr. Salvadori is even more so and also Italian.

To appreciate my brother's book, you must understand one thing about him. He is crazy.

As he states in the book's introduction, "The student can solve more interesting and complicated problems by combining numerical analysis and programming, thus becoming more conscious of the power of modern computation procedures."

This may seem vague to the reader, but I think once you realize that my borther is crazy, it doesn't bother you much.

I do not wish to reveal the plot of my brother's book, but the reader gets a hint of what is coming on page 74.

Discussing "Gauss' Elimination Methods," Mr. McCormick writes: "The unknown xl is first eliminated from $n - 1$ equations by dividing the first equation by $a - 11$ and by subtracting third equation multiplied by $a - 11$ (i equals 2, 3..., n), from the remaining $n - 1$ equations. (The elimination process is thus seen to be identical with pivotal condensation.)"

This is something people have whispered about for years, but never has it been discussed so openly and frankly. The book holds this pace constantly. For me the high point is on page 150, when Gauss, after many adventures, turns to his exotic companion and says coldly, "With Nl previously defined as $N - 1 + 1$, $A(Nl)$ must be used rather than the improperly subscripted variable $A(n - 1 + 1)$, because the most general form of a subscript in Fortran is a constant times the subscript, plus a constant, e.g., $6 * 1$ 219."

I think it's the asterisk that makes this passage so pregnant with meaning. I poured myself a glass of bourbon and stared out the window at the moon for an hour after reading it.

Because my brother wrote some parts and Mr. Salvadori other parts, it's sometimes hard to tell whose prose you are reading. But I know that on page 178, it's my brother writing: "Notice that the variables appearing in the subprogram definition are 'dummy variables.' Thus, Real (1) in Program 4.2 is the same variable as XRL in Program 4.1, REAL (2) is the same as XR2, and RT111is the same as XR2, and RT11 is the same as X1."

This passage has the unmistakable cold, hard, clear pitch of a McCormick.

In contrast, take this, by Salvadori I'm sure: "Iterative procedures are usually terminated when two successive iterates agree to a specified number of significant figures. The rather complicated calculations determining this 'iteration cut-off' are omitted here so as to not obscure the basic simplicity of the iterative process."

I don't think my brother wrote this passage, with its blatantly pragmatic tone, because although he is crazy, he is no extremist.

This book is not meant for children, unless they're crazy, but it is just the sort of reading for passing the time if your parachute fails to open.

My brother asked my frank, critical opinion, and I told him he and Salvadori weren't in the same class with Joyce yet, but at least they wrote nice clean stuff that you weren't ashamed to be seen reading in the psychiatrist's office.

Bernard McCormick
Delaware County Times
Sept. 9, 1964

What is Electricity?

Today's scientific question is: What in the world is electricity?

And where does it go after it leaves the toaster?

Here is a simple experiment that will teach you an important electrical lesson: On a cool, dry day, scuff your feet along a carpet, then reach you hand into a friend's mouth and touch one of his dental fillings. Did you notice how your friend twitched violently and cried out in pain? This teaches us that electricity can be a very powerful force, but we must never use it to hurt others unless we need to learn an important electrical lesson.

It also teaches us how an electrical circuit works. When you scuffed your feet, you picked up batches of "electrons," very small objects that carpet manufacturers weave into carpets so they will attract dirt. The electrons travel through your bloodstream and collect in your finger, where they form a spark that leaps to your friend's filling, then travels down to his feet and back into the carpet, thus completing the circuit.

Amazing Electronic Fact: If you scuffed your feet long enough without touching anything, you would build up so many electrons that your finger would explode!   But this is nothing to worry about unless you have carpeting.

Although we modern persons tend to take our electric lights, radios, mixers, etc. for granted, hundreds of years ago people did not have any of these things, which is just as well because there was no place to plug them in.   Then along came the first Electrical Pioneer, Benjamin Franklin, who flew a kite in a lightning storm and received a serious electrical shock. This proved that lightning was powered by the same force as carpets, but it also damaged Franklin's brain so severely that he started speaking only in incomprehensible maxims, such as "A penny saved is a penny earned."   Eventually he had to be given a job running the post office.

After Franklin came a herd of Electrical Pioneers whose names have become part of our electrical terminology: Myron Volt, Mary Louise Amp, James Watt, Bob Transformer, etc. These pioneers conducted many important electrical experiments.   For example, in 1780 Luigi Galvani discovered (this is the truth) that when he attached two different kinds of metals to the leg of a frog, an electrical current developed and the frog's leg kicked, even though it was no longer attached to the frog, which was dead anyway. Galvani's discovery led to enormous advances in the field of amphibian medicine. Today, skilled veterinary surgeons can take a frog that has been seriously injured or killed, implant pieces of metal in its muscles, and watch it hop back into the pond just like a normal frog, except for the fact that it sinks like a stone.

But the greatest electrical pioneer of them all was Thomas Edison, who was a brilliant inventor despite the fact that he had little formal education and lived in New Jersey. Edison's first major invention in 1877, was the phonograph, which could soon be found in thousands of American homes, where it basically sat until 1923, when the record was invented. But Edison's greatest achievement came in 1879, when he invented the electric company. Edison's design was a brilliant adaptation of the simple electrical circuit: The electric company sends electricity through a wire to a customer, then immediately gets the electricity back through another wire, then (this is the brilliant part) sends it right back to the customer again.

This means than an electric company can sell a customer the same batch of electricity thousands of times a day and never get caught, since very few customers take the time to examine their electricity closely. In fact, the last year any new electricity was generated in the United States was 1937; the electric companies have been merely reselling it ever since, which is why they have so much free time to apply for rate increases.

Today, thanks to men like Edison and Franklin, and frogs like Galvani's, we receive almost unlimited benefits from electricity. For example, in the past decade scientists developed the laser, an electronic appliance so powerful that it can vaporize a bulldozer 2000 yards away, yet so precise that doctors can use it to perform delicate operations to the human eyeball, provided they remember to change the power setting from "Vaporize Bulldozer" to "Delicate."

Dave Barry
Miami Herald, 1982

A Short Life

Among the many notable achievements during the celebrations of Einstein's 100th birthday, the discovery of his coworker in Berlin, S. B. Preuss, was one of those receiving rather little publicity. This may have been caused by the accidental nature of the discovery. In a section of their review article about cosmology entitled "Historical Comments," the authors state: "The discovery...of Hubble's law...also led Einstein to immediately reject the notorious cosmological term (Einstein and Preuss, 1931)."

The curious reader who has followed Einstein's life story and knows of his collaborations with M. Grossman, J. Grommer, and W. Mayer (to name a few), but who has never heard of S. B. Preuss, eagerly turns to the reference given. It is: A. Einstein and Preuss, S. B. (1931), *Akad. Wiss.*, 235. Surely the *Akad. Wiss.* must be the Berlin academy. Handily enough for those without access to the originals, Einstein's reports to the Berlin academy have been reproduced on the occasion of the 1979 celebrations and can be bought with little cost. A glance at page 235 of the 1931 volume of the *Sitzungsberichte der Preussischen Akademie der Wissenchaften, Phys-Math Klasse* (Reports of the meetings of the Prussian Academy of Science) reveals the working of a creative mind. Not inclined to check the primary sources, the heedless citation quoter nevertheless is able to compensate for a considerable loss of information. Let us look at the following sequence of references to the author and journal in question:

Einstein, A. (1931). *Sitzungsber. Preuss. Akad. Wiss....*
A. Einstein, 1931, *Sitzgsber. Preuss. Akad. Wiss....*
A. Einstein, *Sitzber. Preuss. Akad. Wiss...(1931)*
A. Einstein (1931) *Sber. preuss. Akad. Wiss....*
Einstein, A., 1931, S. B. Preuss. *Akad. Wiss....*
A. Einstein, S. B. Preuss, *Akad. Wiss.*, 1931.
A. Einstein, S. B. Preuss, *Akad. Wiss.* (1931)....
A. Einstein and Preuss, S. B. (1931) *Akad. Wiss....*

Thus, it turns out that the birth and death of S. B. Preuss occurred within such a very short time span that any scientific endeavors attempted could come to nothing. One hopes that this will be noticed by the public, including the people producing the citation index. Otherwise, I fear, lest, in a generation or two, a young historian of science might apply for a grant to uncover more details from the brief, but not entirely joyless, life of S. B. Preuss.

Hubert F. Goenner
Physics Today
May 1982

Fun Research

How To Be a Project Leader—
Nine Helpful Hints

The purpose of this lecture is to give you practical, helpful hints on how to be a successful Project Leader. You will be told how to start a project, how to direct it, how to write the report, and how to go on the road and sell the project report.

I have titled this talk "Nine Helpful Hints." Some of you may wonder why there are only nine helpful hints when there are ten common pitfalls. Well, you just can't think of everything.

I. State the Problem

A judicious statement of the problem is the most important single factor in originating a good systems analysis. It should not be too specific, for several reasons. If you make the problem too specific, you have no latitude. But more important—if you are too specific, you may find that the idea will not be approved. If you are obscure, no one will understand what you are going to do, and you will be given wide latitude in doing it.

Great wisdom must be exercised in setting the scope of the project. If you limit the scope to what you think can be accomplished in a reasonable period of time, your superiors will more than likely say, "That's a good idea. Go ahead and do it." And you will never get to be a Project Leader (PL).

But you must not be too ambitious. If you ask for too many people, you may well find your superior (or someone of equal stature with more pull) latching on to the project and you will end up an Assistant Project Leader (APL). The plight of an Assistant Project Leader is a sorry one, indeed.

These are the only things to worry about in *Stating the Problem*. You will not have to stick to the course of action you outline in the statement. The incongruities of systems analysis are such that once the problem has been approved and you have been designated Project Leader, you can start the ball rolling down any road you choose.

II. Assemble a Team

The next step is to get some people working for you. Not just a few people. Get a lot of them. Remember that there is no fun and no glory in directing (distracting) only a few people.

First, designate an Assistant Project Leader. Great care should be used in selecting this person. Remember, he or she must run and supervise the project and is your Chief of Staff. Give your Assistant complete freedom. Let APL worry about the Project. You must meditate on the Big Picture.

Next, select your technical team. You may find it desirable to choose people with some knowledge of the problem.

You should include an economist. Someone has said that economics is only common sense made difficult. This is true and thus you will need at least one and preferably two economists on your team. (If you have two, they can argue with each other.) They are good at taking straightforward data and putting it into the language of systems analysis. They use such terms as optimization, suboptimization, allocation of resources, marginal utility, and the like.

Then you need a social scientist. Having a social scientist on your team will add a certain amount of prestige. This person will be indispensable when it comes to writing the report. One good social scientist can contribute a hundred pages of your report without even knowing what the problem is.

Be sure to include mathematicians—even if you agree with Plato, who said, "I have never met a mathematician capable of reasoning." Any systems analysis worth its salt must have several appendices full of equations, and this is where mathematicians come in handy. If the mathematicians cannot put all the information into equations, they will recommend war-gaming it, in which case you will have more and more people working for you.

Don't forget a physicist. Physics is a very proper and popular science. Physicists also know about equations. Some of them know equations the mathematicians don't know, so you are providing yourself with added protection. You will find a physicist indispensable when you have a conference, for it is typical of their breed that they will debate vigorously on any subject.

To this essential cadre you may add consultants on almost any subject—psychologists, engineers, and others. Just be sure that you get a well-rounded group. Add more people as the study progresses.

III. Have Conferences

Nothing contributes as much as conferences to the appearance of great activity in a systems analysis. At least three conferences a week is a goal to strive toward.

Early conferences may clarify the problem, as well as reveal ways to expand the study and thus increase the number of people on your team.

Conferences may also keep you from having to work on the problem. Some Project Leaders have found it possible to discuss a project until some other agency has written a report on the subject. This is not as alarming as it may seem, for then you can (1) modify their work, correct their assumptions, and turn out a first-rate report; or (2) start your team working on another project that you know more about; or (3) report on their report.

IV. Gather Data

Before you can tackle any problem successfully, you must have facts—lots of facts.

An inexperienced Project Leader will suggest that each individual gather data in his or her own particular field. Wiser heads know that such an approach stifles originality, so they encourage the more imaginative team members to investigate fields other than their own. An unexpected payoff may be the discovery that an economist is really a first-rate (but frustrated) mathematician, or that an electronics engineer is really a social scientist at heart.

Your success as a Project Leader may depend on decisions concerning the amount of details to be obtained.

Ordinarily, one should not worry about this in the early stages of the project; you can always "wash out" the details by making assumptions when the Project bogs down or loses momentum—when it looks as if it shouldn't have been started in the first place. The wise Project Leader will then search for some obscure and little-understood detail, and gravely announce that work cannot proceed until this point has been cleared up.   Not only does this gambit show the awareness of the Leader, it also gets our PL off the hook until somebody on the team has suggested the next step in the project. This latter endeavor may tie up two or three more people, thus increasing the prestige of the Project Leader.

One other item: Find some little-known references, preferably in foreign-language journals or reports. Nothing gives a technical report as much prestige as references like: Wagner, A., Theorie und Beobachtung der periodischen Gebirgswinde, *Beitr. Geophys.* **52**, 408-459 (1938).

Attention to these details separates the sheep from the goats as far as Project Leaders are concerned.

V. Take Trips

One of the advantages of being a Project Leader is that you can take trips. Trips should begin as soon as you've had enough conferences to determine which team members make the best companions. A party of four or five is ideal—the exact number depends on whether you prefer bridge or poker. Company or staff cars at headquarters can handle either four or five passengers comfortably, and when you travel by taxi, you can divy up the fares.

Any systems analysis should envisage at least four types of trips: (a) to Headquarters, to get the "big picture" from Top Management; (b) to a major divisional site, where you will find that Top Management doesn't understand the problem; (c) to an overseas subsidiary, to get the views of "the troops in the field." Care should be taken to visit Europe in the summer and southern hemisphere sites in winter; (d) to contractors, to get the feel of the "hardware and software of tomorrow."

When you visit other organizations, make sure that you have appointments with the Chief Executive Officer. Then you can quiet your opponents with "When I talked to their CEO in December," or "That may well be, but when I visited the commanders in the field in June—."

When you go to Europe, take along somebody who can read menus and street signs. If you are fortunate enough to have a little capital, you should bring some less favored associate who will let you use his or her duty-free exemption when you return to the States.

V. Use Your Tools

After the preliminary round of conferences, after fact-finding trips are completed, and after a sufficient amount of raw data has been obtained, it is usually advisable to make some calculations.

A *must* for a systems analysis is a high-speed computer, such as MANIAC, Cray, NATPAC, Amdahl, or GENIAC. According to the manufacturer, the Geniac (twenty dollars) may also be used during lulls to solve such puzzles as "The Space Ship Airlock," "The Uranium Shipment and the Space Pirates," "The Two Jealous Wives," and "The Fox, Hen, Corn, and Hired Man." Few people will argue with a computer, feeling that it combines the qualities of a sorcerer with those of a slightly mad (but competent) scientist.

There are other tools available. More sophisticated studies involve somewhat more abstruse tools, such as combinatorial analysis, the Lagrange multiplier method, Dilworth's theorem, etc.

Statistical uncertainty is an important component of any systems analysis, but this can be handled in a number of ways. Studies have employed coins, cards, dice, or roulette wheels to solve statistical uncertainty problems. These tools are quite satisfactory if used properly. However, more than one system study has literally distintegrated at this point because team members have used the tools in the ill-conceived hope of private gain.

Some analyses handle uncertainty through use of a recent Rand book, *A Million Random Digits with 100,000 Normal Deviates*. I have never used the book because I don't consider deviates to be normal. (One can't be too careful when one has a Top Secret clearance.)

VII. Observe Security

(It is unfortunate that this section, by far the most important, cannot be released at this time because of security restrictions.—Ed.)

VIII. Write the Report

Now we come to the most onerous task in any systems analysis—writing the report. A few basic principles may be noted here.

If you have done a good job in your study, make the report as short as possible. If you have any doubts about your study, increase the number of pages of your report in direct proportion to your anxiety. On this you may need the help of your economist and social scientist.

Before starting to write the report,* call a conference and remind your team that:

(a) Data and calculations should be checked against the reports of other agencies to avoid direct comparison and possible embarrassment.

(b) When the data for curves and graphs are uncertain, the vertical scale should be placed far to the left of the page, so that when the report is bound, the reader must unfasten the pages to read the curve. Few readers are so curious.

(c) In many cases, inadequate or incomplete data may be disguised or concealed by assumptions. Let us say that you are comparing the effectiveness of Air Force weapons against people in foxholes. This is a difficult task, but the problem can be handled easily. You can eliminate high explosives from the

*All reports should be called 'Interim' reports.

study by assuming thlat the required logistics support would make the use of such weapons infeasible. You can eliminate biological and chemical agents by assuming that, because they were not used in World War II, they will not be used in the future. You can eliminate napalm and rockets by assuming that the target is beyond the range of fighter aircraft. Thus you are then able to resolve the problem into one for which you know the solution.

In preparing the conclusions and recommendations, you must not be too specific. You will find that the best conclusions are those that could have been made with intuitive reasoning prior to the study. The best recommendations are those that require: (a) a small increase in personnel, equipment, or facilities if the sponsoring agency must pay for the increase from its own budget, or (b) a large increase in those items if the money is to come from somewhere else. Of course, your recommendations must point out that continued work on the analysis is necessary.

A Project Leader has two choices in designating the author or authors of the report. If you have doubts about the value or validity of the report, you can give your associates equal (or even higher) billing on the title page with you. This is the "share-the-blame" principle. If you think the report is good, but not too good, you can designate yourself as Editor, giving a prominent place on the title page to your team members. If you think the report is really good, you may list yourself as Author, and relegate your associates to an obscure section of the report entitled "Acknowledgments."

IX. Go on the Road

We now come to the briefing tour. There are two schools of thought on timing. If the tour *precedes* the final report, you can incorporate worthwhile comments in the report when you write it. (However, some timorous Project Leaders have been so unnerved by the briefing tour that they *never* wrote a final report.)

There are inherent advantages in having the tour *after* the report is written. Then you can give a short briefing, covering only the highlights of the study. Questions can be parried by holding up a copy of the report and saying: "The answer is in one of the appendices here."

There are also two schools of thought on how to use your team on the briefing tour. You may act as master of ceremonies and let your associates make the presentation. This "Kindly Senior Statesman" technique impresses your audience with your tolerance, understanding, and patience. You leave the impression of knowing all the answers, but of letting your associates answer to get experience.

If your study is good, you may wish to give the briefing yourself. Then, you can use the members of your team to change charts, to look attentive, and to help you with the more difficult questions.

In conclusion, I can only wish you well, with the hope that these "Nine Helpful Hints" may guide you on your way toward becoming a successful Project Leader.

Harvey Lynn, Jr.
Operations Research
Vol. 4, pp. 484-488 (1956)

An Application of Gamesmanship in Science

Faced with an apparently interminable stack of examination papers that threaten to take up all the available research time for days to come, the scientist who also teaches must often wonder if there is not some easier way of "moving along" in the profession. Many expedients have been suggested, but few of them are really workable. Grading papers by throwing them downstairs, marking at random, and adjusting the grades of those who complain, or simply not giving examination, are devices that come readily to mind. All of these have certain drawbacks—and furthermore, they all fall short of the real objective of "moving along." They merely speed the professor back into the laboratory to be once again faced with an apparently interminable stack of specimens, miscellaneous

undigested data, or incompleted experiments. If the real objective of "moving along" is to be attained, the individual must sys-
tematically pursue this goal. The scientist must not progress by
a series of jerks and bounds, but press steadily forward. Pause
not to look back, as Professor Challenger warns, but ever forward toward our "glorious goal."

What we need, if we have no philosophy to guide us, is a
system. Fortunately for those of us who are not so gifted as to
be able to think out our own, an almost infallible set of techniques has been suggested by that peer of English humorists,
Stephen Potter.   In his book, *The Theory and Practice of
Gamesmanship,* which every thinking scientist should read,
Potter outlines the general principles of his newly defined, but
far from new, system. It may be encapsulated as the art of winning, or in our contest "moving along," without actually indulging in engañando, as they say in Mexico.

It is, of course, impossible in the short space of a single
article to expound all of the possible uses of gamesmanship in
science. We can only hope to outline the barest principles and
hope that the readers* can develop them along lines fitting their
own needs. We, therefore, content ourselves with brief discussions of two areas in which gamesmanship is immediately
applicable to the situation of the college or university professor,
but which at the same time clearly illustrate most of the
pertinent principles.

Publication

The emphasis upon publication, as a measure of academic
worth, has become so great in recent years that the professor
should seriously consider the problem. No fixed scale correlating with the probability of advancement and promotion has as
yet been established, blut some tentative parameters may be set
up. In large competitiive institutions, an average of twelve pap--

*We debated some time whether this article should be offered to a
publication that may be availabe to graduate students before they have been
granted their final degrees and are thus removed from direct competition.
Students, after all, must be kept absolutely in the dark about what actually
goes on until the last possible moment before they are thrust out into the
marketplace.

ers a year probably represents a favorable balance for the individual. In smaller schools, less publication and more attention to other details may be indicated. Quality of published work must be considered in connection with the desired ends. If permanent residence at a particular place is contemplated, the products should probably be of a somewhat higher calibre than if frequent and recurrent changes are appealing. Most deans apparently read only the titles; others simply count, etc.[†]

Even the harried instructor of undergraduates can usually find time to grind out a few notes each year, but wisdom suggests that these should not be published in the journal of any society that does not publish separate abstracts. This practice will quickly and automatically double the instructor's apparent productivity. Great care should always be taken to see that there's some change in title before final publication. This confuses bibliographers no end and assures the citation of both abstracts and final paper appears in later works. Never include too much in any one abstract, because you will find it far more productive to split up the final results into a whole series of abstracts and papers.

To maintain a really rigid production schedule, however, additional methods are usually needed. Joint publication with graduate student seems an obvious answer, but great care and tact must be used in practice. In American universities, the students, particularly the better ones, seem overly wary of any attempt to join them in authorship. If you insist upon your prerogatives, you may find that they are actually concealing their results from you until it is too late. Also the use of too much force may so prejudice your later relationships that you will be cut off from future revenue in the form of hints about good lines of experimentation, etc.

If you are content with junior authorship, which in terms of gamesmanship is an acceptance of defeat, several rather mild gambits may be worked out. The "You can see what a fix I'm in here, Jones," approach may be enought to appeal to the filial

[†]This quantitative approach to academic values is firmly rooted in Pythagorean philosophy. English deans probably have an advantage over those in America since they have the possibility of correlating the number of letters after the man's name with the number of publications. Heavy covers for reprints are also suggested in case weighing may be resorted to.

instincts of many. Wear your old suit that day and don't be too careful about brushing the talcum out of your sideburns. Other scientists will be less malleable and more rigorous methods will have to be applied if you are to be successful in getting on a paper's list of authors.

Those who are fortunate enought to have more than one graduate student at the same time are doubly blessed. It is extraordinarily easy to keep a group at just the right degree of pique with each other so that you can abstract the pertinent results from two or more projects and attach yourself as senior author. The junior authors will seldom discover that you contributed absolutely nothing. They will believe that the ideas they did not have came from you.

With only a single student, the problem is more difficult but far from hopeless. Be sure to insist that the student submit all manuscripts intended for publication to you. Old data on experiments or observations you made years ago can always be brought into the manuscript. Their real relevance is immaterial. Any such material available can be incorporated and then diligently edited out later. In the process of editing, these simple rules of thumb may be followed: (1) Compound all simple sentences: (2) Rewrite all compound sentences as simple ones: (3) Change "probably" to "it seems," and "it seems" to "probably," and (4) Rearrange all papagraphs several times being sure to remember the original order if you do not trust your own finesse in such matters. After several such revisions, the suggestion that you be made senior author so you can help complete the paper, is usually acceptable. If not, do not give up before trying a few more revisions. Once the goal of senior authorship has been achieved, the original format and style can be restored. The only difficulty with this technique is that, unless you have a secretary, it may take longer than doing the work yourself.

A final suggestion in regard to publication is to begin a series. Others have pointed out that there is no need to start with number one. In fact, a title such as, "Studies on the Regeneration of Cytochrome X in the Decapitated Chick XX" is in itself a thing of beauty to add to your bibliography. The addition of "Further" to the title seems to us redundant, however.

Impressing the Student

In some colleges and universities, the pernicious idea that a professor ought also to teach has become prevalent. In some extreme cases, promotion and pay are in part based on the teacher's success. The worst possible situation arises when some sort of student poll is used as a basis for judging teaching. We heartily condemn any such practice, for our collective observations clearly indicate that the students are invariably wrong in selecting "the most popular professor" or "the teacher who influenced our thinking most." No amount of condemnation will change some such situations, however, so a straightforward recognition of their existence and suitable countermeasures are necessary.

Space permits us to mention only a few possible approaches to this problem of impressing the student. One easily applied technique is evident, however. In any course, be sure to select a text of imposing format and price, but at least ten years out of date. Then, prepare all lectures from the most-up-to-date texts available; but be careful to keep these checked out of the library, and never mention their existence to the students. This leaves the indelible impression that you are really "on your feet," "right in there pitching," and a "real hot shot."

Those who feel that the use of gamesmanship against the students is unethical should remember that even in those schools with the smoothest honor systems, it is said that the professors have the "honor" and the students the "system." We are merely suggesting that the professor adopt a system also. Anyway, it might actually improve your teaching if you read the latest textbooks.

With these few hints on the basic principles and application of gamesmanship, we leave you to your own devices. Now, back to the bluebooks...

F. N. Young and Sears Crowell
AIBS Bulletin
Vol. 6, p. 13 (1956)

Sciencemanship

The latest book on "how to be a good scientist" is by Prof. D. J. Ingle of Chicago, USA, and has the title *Principles of Biological and Medical Research*. At the end I found myself wondering why it had so little to add to the admirable earlier studies of Beveridge *(The Art of Scientific Investigation)* and of Wilson *(An Introduction to Scientific Research)*. Perhaps it is not so easy to break fresh gound in this field. However, I am not entirely convinced of this.

Murphy's Law

I think the deepest and most durable impression that the researcher sooner or later gains is just how unexpectedly, how unjustly, how distressingly difficult it seems to be to discover or prove anything at all. The research worker would be spared much early perplexity if his or her formal instruction included a sound treatise on Murphy's Law.

This important Law is described by H. B. Brous Jr. in the September issue of *Astounding Science Fiction* as stating: "If anything can go wrong, it will." Researchers among my readers will instantly recognize the truth and generality of this Law, even if they have not previously come across its verbal formulation. But having recognized it, what to do about it? Ingle's book, in common with those of Beveridge and of Wilson, has no definite proposals to make.

Here, then, is a suggestion, offered as a stimulus to others interested in this uncharted territory. The moment to take account of Murphy's Law is clearly when you are planning a new investigation. You have worked out how much material will theoretically give you the required amount of information. We will call this the theoretical estimate, x. Here x may be the number of rats to be treated or the acres to be sown or soil samples to be collected, and so on. You then attempt to make rational allowance for all the things that might go wrong. Although judging any specified mishap to be highly improbable, you might yet consider that the joint effect of all the improbable mishaps might amount to, say, a possible 30% wastage. You therefore decide to budget for 1.43 times the theoretical estimate (after 30% wastage, $1.43x$ becomes x), and the multiplier you use (in this case 1.43) I call the Rational Multiplier, R.

$$M = R^2$$

It is at this stage that we usually finalize our plans, and live to regret it. It turns out that although some of the possible hazards did not materialize, we had forgotten that a proportion of the rats might have fatal convulsions on hearing a whistling kettle, and that a colleague might mistake some of the clearly labeled organs stored in the refrigerator for goldfish food, and act accordingly. It is before any of this happens that Murphy should be consulted. Having quizzically surveyed the wreckage of many an experiment, I assert that the needed prophylactic lies in the use of the Murpheian multiplier, M, in place of R, to which it is related by the simple expression $M = R^2$. In our hypothetical case, supposing that the inexperienced (entirely theoretical) investigator would procure 100 rats from the dealer or animal house, the "rational" investigator would procure 143, but Murphy would procure 203.

The expression $M = R^2$ rests on more than empiricism. It was derived, with the aid of my colleague, Anne McLaren, from certain theoretical considerations. These involve the idea that the Rational Mulitplier depends on the number of discrete risk-bearing operations into which the total experiment can be broken down. If the rationally foreseeable risk attached to each of these is assumed to be accompanied by an independent, unforeseen, Murpheian risk of equal magnitude, then the above equation follows. Of course, the assumption is crudely approximate. But it is a beginning.

Necessity of Idleness

Beveridge has emphasized the need of the research man to lie fallow for periods of time, and quotes J. Pierpont Morgan as saying, "I can do a year's work in nine months, but not in twelve months." Unfortunately, he offers no concrete suggestions. A former colleague at one time installed a camp bed in his lab so that he could lie down when he felt tired or lazy. His

Department Head disapproved, but I think that the idea is interesting.

An allied problem is the Visitor Menace. I knew a famous man of science who, when a self-invited visitor was in the offing, would retire to the cloakroom. He took with him his papers and books, and emerged only when the "all clear" was sounded. I find no reference to the cloakroom maneuvre in Ingle's book, and Beveridge and Wilson are also without practical recommendations.

The visitor menace is an expression of a general, and truly paralyzing, affliction that overtakes most researchers in their mature years. This is the Ernestness of being Important. Ingle says, "The early years in the laboratory are the golden years for many scientists. After one becomes known, the volume of mail, telephone calls, number of visitors, organizational activities, including committees by the dozens, and demands for lectures, reviews, and community activity grow insidiously and will destroy the creativity of the scientist if unopposed." But how to oppose them?

Five Principles

If only to start the ball rolling, here are five principles of evasion, not yet tried and tested, but perhaps deserving of trial.

1. No committees
2. No refereeing
3. No editing
4. No book-reviewing
5. No invited papers

Special dispensation can possibly be granted for anything for which the hard-up researcher can get sufficiently well paid (for example, reviewing Ingle for *Discovery*). The fifth is the least obvious, but rather interesting. I added it recently when I had been going through my collection of reprints of the scientific papers of others to discard those that I felt I could do without. At the end, I found to my surprise that my reject pile contained a high proportion of papers that had been delivered by invitation to some conference or symposium.

The clue probably lies in the recipe that one tends to follow for putting together an invited paper for a special occasion. The recipe is hash and waffle. By "hash" I really mean *re*hash or results that have in the main already been published elsewhere. The concoction can be diverting and informative for one's listeners. But it seem that hash and waffle is a dish that does not keep.

The explosive expansion today of almost every sector of the scientific front makes a vital necessity of any and every means of keeping scientific workers in touch with each other, and with the latest advances in their own and neighboring fields. The scientist who helps to perform this service richly deserves the gratitude and admiration of fellow investigators. But let our good scientist not imagine that an original and lasting contribution to knowledge is made thereby. If *this* happens to be the scientist's ambition, there can be no compromise. Such a scientist must be perpared ruthlessly to disembarrass his or her thoughts and timetable from every preoccupation other than the central quest.

Chap Rotation

I once worked in an applied research outfit that among other peculiar practices, operated a sort of rotation of crops, or, rather, rotation of chaps. Once in every while—I do not now recall whether this was once in six or seven or eight weeks— each scientist was banished to a small room for a week, in which the only duty was to sit and muse. No one asked at the end of the week, "Did you have any bright ideas?" for this in itself might damp the muse. Our scientist was only asked to abstain from all routine work during that week. In exchange, the exile had arbitrary powers to commandeer any of the outfit's equipment or labor force to test any bright ideas that popped up.

Some heads of research teams may look askance at this scheme. To those who are tempted to try it in their lab, I should emphasize the following. It must be made very clear that the exile who spends an apparently barren week with his feet on the table reading the comics gains the same merit in the eyes of the team and its leader as the one who emerges to suggest six new experiments and a modification of the Second Law of Thermodynamics. Otherwise the whole point is lost.

Browsing

The rotation of chaps is only one of many possible devices for recharging the researchers mental batteries. The necessity of recharging is eloquently stated by Kursanov, employing a different metaphor: "A scientist is not a balloon, to reach a certain height and remain there for a long while on account of the material it was once filled with. Friend scientist is "heavier than air," more like an aeroplane that has to keep going to maintain its height or to climb." It is well known that height is on average not maintained. Beveridge cites Lehman's figures for output at different ages. Taking the decade of life 30–39 as 100, the output for the years 20–29 was 30%–40%; for 40–49, 75% for 50–59, about 30%. Assuming that slow start is caused by lack of knowledge and experience, is the later decline entirely caused by biological aging? I think not. The features of a young scientist's life that tend to disappear with time and that may be important at once occur to me. One is browsing and the other is fairly frequent change of work and surroundings.

What senior scientist can be found sitting all day in the library looking through research periodicals because there is nothing else particularly to do? And what research student does not from time to time do just this? As for change of work, Beveridge mentions the case of Ostwald, who successfully rejuvenated his mind by this means when he was over fifty years of age. In this connection, a proposal made by Kursanov's countryman, the nuclear physicist Peter Kapitsa, deserves attention. Kapitsa intends his suggestion for adoption in Russia, but there seems no obvious reason why it should not be applied more widely.

Combat Forces

His idea is the setting up of *ad hoc* "mobile combat forces," each to be regarded "not as a permanent institution but as one set up to tackle a given problem over a period of months or years." Such a force would consist of scientists drawn from a number of different specialties, each with some special angle on the problem to be solved. After the successful solution of the problem, the combat force would be dissolved and its members would return to the permanent departments or institutes from which they had been recruited, or some of them might join new combat forces.

Something like this in fact occurred in Britian during the war, but with a measure of compulsion inadmissible in peacetime. Apart from the gain in efficiency, I see a valuable psychological advantage in such a scheme. It would enable even the senior researcher to reverse a trend toward stagnation, for the scientific mind is more like medicine than a wine: it should be well shaken before use.

Many readers will have other, and better, suggestions than those aired here. But enough, I think, has been said to show how many and how inviting are the paths Ingle has failed to tread.

Donald Michie
Discovery
June 1959, p. 259

Bumper Stickers

The Astronomical Society of the Pacific, San Francisco, has contributed to the exhibitionist literary genre, the bumper sticker, making the following available:

Astronomy is Looking Up
Supernovae Are A Blast
Black Holes Are Out of Sight
I Watch Heavenly Bodies
Turn Off Your Lights,
Turn On to Astronomy
The Big Bang is A Naked Singularity
Pulsars Turn Me On—Fast
Red Giants Aren't So Hot

Blue Stragglers Live Forever
Inside Outer Space—Visit Your Planetarium
Astronomers Do It At Night
Interstellar Matter Is A Gas
Forbidden Lines Are Exciting
Quasars Are Far Out

Astronomy Society of the Pacific

The Twelve Coin Problem

Professor Felix Fiddlesticks
Is always up to foolish tricks.
His latest game is to collect
All fakes and duds he can detect,
And counterfeits. He says it is the truth
Forgers are not what they were in his youth.

One day, by high ambition lit,
He forged a perfect threepenny bit.
It was about as good a fake
As anyone could ever make.
His pride at this success he couldn't smother,
He rushed off home to show it to his mother.

"Oh, Mother, see what I can do,"
And from his pocket he withdrew,
Not, as he thought, one three penny bit,
But twelve—alas! For all his wit,
The counterfeit he just could not locate
By sight, but he knew it differed in its weight.

"Oh, clever Felix Fiddlesticks,"
His mother said,"you're in a fix.
The spurious threepence, can you state,
Is light, or is it overweight?"
"I can't remember." "Here's a balance, see,
Go find the counterfeit in weighings three."

Solution to the Twelve Coin Problem

F set the coins out in a row
    And chalked on each a letter, so,
To form the words: "I AM NOT LICKED"
    (An idea in his brain had clicked).
A bold man must he be who thinks he licks
    Our wonderful Professor Fiddlesticks.
And now his mother he'll enjoin:

	Coins put on left-handed side	Coins put on right-handed side
1st weighing	MA, DO	LIKE
2nd weighing	ME TO	FIND
3rd weighing	FAKE	COIN

By weighing thus, he can detect
    The spurious coin by its effect;
And more than that, with confidence he'll state
    Whether the dud is light or overweight.

For instance, should the dud be L
    And heavy, here's the way to tell:
First weighing, down the right must come;
    The others, equilbrium.
Each coin can thus be tested—or perhaps
    F left the dud behind, a frequent lapse.

Such cases number twenty-five,
    And F's scheme we so contrive
No two agree in their effect,
    As is with pen and patience checked:
And so the dud is found. Be as it may
    It only goes to show Crime Does Not Pay.

Blanche Descartes
Eureka (No. 13)
October, 1950

Nicholas Copernicus and
the Inception of Bread-Buttering

Historians have classically emphasized social, economic, and cultural factors as having the greatest impact upon the development of dietary habits. This essay will elaborate upon certain political and military circumstances and their relevance to the inception of buttered bread.

During the first quarter of the 16th century, Ermland was the scene of frequent and terrible devastation. Bordering upon the Gulf of Danzig, Ermland was one of the four dioceses into which Prussia had been divided for purposes of ecclessiastical government. Its bishop, Fabian von Lossainen, was both temporal and spiritual ruler, and his secular responsibilities were more burdensome. A vassal of the King of Poland, the Bishop was frequently engaged in conflict with the Order of Teutonic Knights. The Order had formerly ruled Ermland and periodically launched military expeditions to regain supremacy.

There were three fortified towns in the bishopric. Frauenburg, Heilsburg, and Allenstein. The final attempt of the Teutonic Knights to reassert suzerainty occurred between the years 1519 and 1521. Anticipating that the Knights would lay siege to the castle of Allenstein, the Bishop entrusted its defense to the most able and distinguished Canon of the Cathedral of Frauenburg, Nicholas Copernicus.

Copernicus' immediate task was not especially arduous. Competently garrisoned, Allenstein Castle was a strategic enclave impregnable to any army with such limited resources as those possessed by the Teutonic Knights, and the Canon promptly instituted the necessary precautions. The Knights did attempt to capture the point, but in tacit recognition of the hopelessness of their task, the siege—while prolonged—was otherwise a very relaxed one.

The event directly relevant to this study occurred some six months prior to the lifting of the siege. At that time plague struck within the walls of Allenstein Castle. Copernicus, as commandant, was of course present.

Much of the renowned astronomer's reputation at the time rested upon his skill as a physician. As a youth he had studied medicine at the University of Padua and, although never securing a medical degree, he did acquire a professional proficiency reputed to be well above the level of most of his contemporaries. Since no detailed list of symptoms has survived, we are unable to identify the specific plague which struck Allenstein Castle. From all indications, however it was a moderately serious affair, and we can with certainty attribute much morbidity but only two fatalities to it. Some victims, including Copernicus himself, fortunately suffered relatively slight indispositions.

Copernicus, who at first contented himself with prescribing routine treatments, became disquieted when in many instances the disease either persisted in spite of his therapy or (which was even more disturbing) reappeared in cases which he had regarded as cured. He determined therefore that rather than limit himself to ineffective treatment for the malady, he would search out and define its etiology and pathogenesis.

News of this rather basic approach soon spread beyond the diocese and traveled as far west as Leipzig where it reached the ears of Adolph Buttenadt. Buttenadt was one of the towering figures of his own age who has somehow been lost in the shuffle of history.

Sometime before 1501 Buttenadt determined to study medicine at the University of Padua. There he encountered the young Copernicus, and the two students from the north developed a genuine affection for each other which presisted throughout their lives. It was Buttenadt who served as Copernicus' sponsor at his initiation in the Guild of Apothecaries and Physicians, but despite early parallels, their careers were to take completely different turns.

We are already acquainted with Copernicus' fame as a physician. However, Buttenadt unlike Copernicus did earn a degree in medicine but hardly ever practiced. Instead he dedicated himself to informational and organizational tasks within the very powerful aforementioned physicians guild. He became, in the very highest sense of the word, a professional propagandist.

Unlike his more famous humanist contemporaries, Buttenadt devoted himself more to defense than criticism of the existing order. Possessed of a unique ability to translate the immediate interests and advantages of the medical profession into arguments and slogans that appealed to the more popular prejudices of his day, he rose rapidly within the administrative hierarchy of the guild and by 1517 had become Grand Master of the Apothecaries of Medieval Allemagne (AMA) the official designation of the organization of the Northern European chapters of the Guild of Apothecaries and Physicians (A and P).

The news of Copernicus' research activities disconcerted Buttenadt, and in his capacity as Grand Master he determined to make a site visit to Allenstein Castle. Buttenadt's trip was both arduous and dangerous, but within six weeks he was granted safe conduct through Teutonic Knight lines and into the Allenstein Castle. A meeting with Copernicus took place almost immediately.

After a few moments of exchanging greetings and reminiscenses, the two old friends arrived at essentials. Buttenadt revealed his concern over reports of Copernicus' activities. He hoped that the reports had no basis in fact and prayed to hear them refuted. Copernicus, however, confessed that not only were such reports accurate, but further revealed he had actually discovered the cause of the Allenstein plague and had instituted measures to prevent its recurrence. Buttenadt was astonished and bewildered by this new development. Activities of the sort entered into by Copernicus could, he contended, undermine the entire contemporary ethos of medical science. The obvious fact was that the efforts of a physician—should he be able to discover and contain the *causes* of any particular malady—would inevitably be directed towards nonpatients, individuals who did not solicit and in some instances might not even be desirous of receiving medical attention.

Copernicus remained silent, spellbound by the force of Buttenadt's analysis. The Grand Master was not yet finished, however. Having exhausted professionally oriented arguments, the Grand Master proceeded to enunciate his personal anxieties. Buttenadt did not allow that in attempting good works man necessarily served as a beneficent instrument. Although he could appreciate Copernicus' noble intentions and actions, he also suspected that these might have inadvertently altered Nature's balance and thus redirected the will of God. This latter possibility was the chief basis for Buttenadt's concern. If true, he feared, it could occasion divine retribution.

Although Buttenadt admitted that his fears, if put in operation, might act to suppress originality in medical science, that
was definitely not his intention. He was an advocate of caut-
ious innovation. The problems with which medical science
dealt had acquired a complexity that rendered them no longer fit
for individual insight and judgment, no matter how perceptive.
To be dealt with adequately, they must at some stage be subjected to collective study involving diverse technical skills, specialized knowledge, and organizational viewpoints. Decision
without collective study, Buttenadt concluded, was apt to be
founded on inadequate information and to lack roundness of
judgment.

Copernicus, awed by the force of the Grand Master's
logic, protested that the Allenstein affair hardly justified such
apprehension. To assure Buttenadt his ministerings could have
no significant impact beyond the castle, Copernicus proceeded
to describe the activities leading to his discovery.

Suspecting some correlation between the plague and the
food available at the castle, he had divided the inhabitants into
separate groups each with special diets. Before long it had
become apparent that the small group which had been denied
bread was the only one free from the plague. If it had been
plausible, Copernicus would have simply dispensed with all
use of bread. The increasingly difficult logistical problem of
keeping the besieged castle adequately provisioned, however,
made such a course of action inexpedient. Subsequent investigation revealed a chain of circumstances that enabled Copernicus to implement a more practical remedy.

Upon assuming command of the castle, the Canon's first
order had been to provide for the defoliation of the surrounding
region to maximize logistical difficulties for the sieging forces.
The defoliation policy was, as anticipated, a two-edged sword.
Since the provisions of the castle would also be sorely limited,
Christian charity required that only those Allenstein inhabitants
too old or infirm to provide for themselves be domiciled within
the castle. Thus, all able-bodied nonmilitary residents were required to withdraw from the vicinity. As a result the castle pop-
ulations consisted of either military personnel directly involved
with defense or enfeebled peasants.

The consequent labor shortage necessitated the impressment of the most vigorous of the remaining peasants into paramilitary service.

By far the most difficult function performed by this group was that of occasional waiter. Allenstein military personnel were not permitted to leave their posts except upon completion of a 12-hour tour of duty. Men on duty during dinner hours were required to take their meals while at their posts; these were brought by those peasants pressed into service as stewards. There was no easy access from the kitchens up to the turrets where the greatest number of guards were stationed, and the stewards were required to carry heavy, full-laden trays across a wide courtyard and up a steep and narrow stairway that rose perpendicularly almost 200 feet above the level of the court. They seldom were able to maneuver the entire distance without some minor mishap.

Although seldom ever losing an entire load of food, they quite often dropped separate items. These morsels were invariably retrieved, brushed free of the more obvious contamination, and restored to the serving board, ultimately to be consumed by unwitting military personnel.

The practice would probable not have produced any serious consequences had not the articles of food most frequently dropped been loaves of bread. The bread baked at Allenstein Castle, and then common to all Europe, was a large, coarse, black loaf that could collect substantial amounts of foreign debris without noticeable discoloration or other obvious evidence of contamination. While other articles that fell into the squalid Allenstein courtyard could, in most instances, be easily purged of unsightly extraneous matter, the bread, seldom showing any evidence of its defilement, was summarily restored to the tray. As the diet of all castle inhabitants consisted in large part of this bread, it also, quite incidentally, included a higher than tolerable level of pathogenic flora.

This knowledge of the specific cause of the plague did not in itself suggest the therapy. Copernicus had, as we have already seen, earlier associated the consumption of bread with the cause of the disease and had regarded as unfeasible its elimination from the Allenstein menu.

We have no records contemporary to the event which indicate how Copernicus achieved a final solultion. The autobiography of Gerhard Glickselig, written some 70 years later, however, describes the incident.

Glickselig's suggestion was that the bread loaves be coated with a thin layer of an edible light-colored spread. After such treatment any foreign matter adhering to the fallen loaves would be readily visible and potential victims would be able to protect themselves from contagion by either cleaning off or discarding the contaminated loaf. A decision to coat all bread with a spread of churned cream subsequently resulted in the elimination of the plague from the castle.

Buttenadt, who combined with his practicality an unusual degree of humanitarianism, regarded the localized nature of Copernicus' clinical success as sufficiently mitigating to dispense with any necessity of discipline for this thoughtless and unprofessional behavior. It was then agreed that the incident should be unpublicized; however, should information of the affair leak out, the conferees decided it best to attempt official sanction by extending the concept of the physician's responsibility to an area provisionally labeled *medicina impedienda.*

Despite precautions to secure the secrecy of Copernicus' study, applications of this elemental principle of (what was soon to be called) preventive medicine were quickly adopoted throughout central Europe, with, ironically, Buttenadt himself doing the most to popularize them.

In 1545, two years after Copernicus' death, the Schmalkaldic War erupted. Buttenadt, who had recently been elevated to the post of executive secretary of the Apothecaries and Physicians, was at the time on an inspection tour of guild chapters. Encountering physical destruction and social dislocations throughout Europe, he became increasingly appalled by the backwash of disease and pestilence. Given the prevailing ignorance of any effective techniques of epidemic control, the rapid spread of plague was inevitable, and Buttenadt, whose voice had once cowed the great Copernicus, now desperately broadcasted the details of the late physician's Allenstein cure. As a result the Copernican System of coating bread was rapidly adopted throughout Europe without benefit of the collective judgment for which Buttenadt had once so eloquently argued.

The question whether a collective body would have anticipated the hazards associated with the accompanying rise in the choleserol content in the diet is unanswerable.

The Schmalkaldic War plagues were unlike the one at Allenstein, and there is no conclusive evidence to support the belief that Allenstein techniques in any way retarded the dreaded scourge. Nevertheless, once adopted, the coating practice proved surprisingly durable and eventually integrated into the dietary habits of the vast majority of Europeans.

The method, as an apparent tribute to its most effective proponent, was described as Buttenadting, and it is, of course, in the edited form, *buttering*, that we describe the practice today.

This essay has been an attempt to trace the evolution of one dietary custom and to demonstrate its development as resulting from military and political as distinguished from socioeconomic factors. Even the practice of bread-buttering has not been exhaustively treated. Nowhere, for instance, has an attempt been made to indicate how the current and nonfunctional practice of buttering one side of a single slice was conceived. Neither has the rationale of continuing the custom been considered even after the introduction and wide adoption of white bread. It is hoped, however, that the subject has been explored sufficiently to demonstrate the need for further studies in the area. It is a stimulating but neglected field possessing unusual opportunities for further research that should be actively pursued.

Samuel B. Hand and Arthur S. Kunin
JAMA 214, p. 2312
December 28, 1970

Elephant Sperm

I studied genetics at Indiana University, receiving the PhD in 1952. Early in my graduate student days there, I was a Teaching Assistant and during one semester I supervised a laboratory section for Introductory Zoology, taken in those days by very "green" freshmen. One memorable lab exercise was given over to matters of reproduction, and the Senior TA, under whose guidance and instruction I was obliged to work, asked that I have on hand for study both microscope slides showing fertilization and early development, and living samples of frog eggs and chicken eggs. Things went well enough until, toward the end of the period, and during discussion one student pointedly noted that, "The class has seen everything now except living sperm cells." I asked him to "wait till next week," and dismissed the class. When I consulted the Senior TA for a source of sperm cells, he told me, "I'm looking at one" and that was all. Taking my mentor at his word, I discovered an insurmountable reticence born of modesty and so forth; and yet I was anxious to keep my word as best I could.

Hence I arranged the following "set-up" on the demonstration table just before the lab period the next week. First I filled a 10-liter battery jar about half full of water, then stirred into it a couple of cups of milk. To this I added some 100 or so recently emerged black tadpoles. The class crowded about the preparation in rapt silence as the small animals swam randomly through the murky waters, each appearing and promptly disappearing from view. Some students made notes and sketches, and all dutifully recorded the information I had printed in bold letters on a 3 x 5 in. card, "Elephant Sperm." One young man broke the intense silence to remark to his peers, "I figured they had eyes, *but look at them tiny mouths!*"

Richard W. Siegel

Poetic License!

A magnet hung in a hardware shop,
And all around was a loving crop
Of scissors and needles, nails and knives,
Offering love for all their lives;
But for iron the magnet felt no whim,
Though he charmed iron, it charmed not him;
From needles and nails and knives he'd turn,
For he'd set this love on a silver churn!

And iron and steel expressed surprise,
The needles opened their well-drilled eyes,
The penknives felt "shut-up," no doubt,
The scissors declared themselves "cut out,"
The kettles they boiled with rage, 'tis said,
While every nail went off its head,
And hither and thither began to roam,
Til a hammer came up—and drove them home.

W. S. Gilbert
Patience, Act II

How to Swat Flies

Sir: The potential for houseflies to spread disease has, of course, long been recognized. A fly, having crawled over human or animal feces, may enter the house eventually to alight on exposed food. Usually the fly, attracted by light, whizzes up and down the window. Attempts to swat it dead are usually thwarted since the fly has a high-speed (millisecond) reflex system in its visual brain–motor system so that it responds by taking off at an avoiding angle in response to a moving approaching swat entering its visual field.

In the interest of hygiene I have experimented on the most effective way of swatting. A piece of tissue paper is taken in each hand and the fly approached from the left and the right, keeping the hands equidistant from the fly and moving to and fro slightly then both hands simultaneously pounce. The fly cannot cope with this situation since its central nervous system circuitry is geared to avoid appraching movement in only one part of its visual field at a time. Two simultaneously approaching swats render the fly immobile, for its central nervous system cannot compute at which angle to take off.

E. G. Gray
Nature 304, p. 678
25 August, 1983

The Case of the Floating Matzo Balls

I relate this story with embarrassment—and pride; embarrassment, because it was the only time my classmate Guillaume Osler and I were bested as amateur sleuths, and pride, because we were taught a lesson in detection and humility by none other than my grandmother, Bobeh Leah. Bill (to give him his English name) and I had acquired some repute as amateur detectives even while pursuing our herculean studies in medical school. In earlier memoirs I had detailed Bill's brilliance, his encyclopedic knowledge of the arts and sciences, and his penetrating logic—traits he undoubtedly inherited from his famous grandfather, Sir William Osler, who had preceded us as a medical student at the same school, McGill University. But for all of our combined talents we were outwitted and outclassed by the most unlikely candidate, my Bobeh Leah.

Bobeh ("grandma" in Russian and Yiddish) migrated to Canada from Russia in her mature years and, attacking the New World with zest, acquired a respectable knowledge of English and French. For some years she lived in a small French-Canadian village at the foothills of the Laurentian Mountains.

Among our classmates we had a Donald MaDonald, intelligent enough to make a good physician, but lacking in diligence and dedication. He was caught up in high living, but he had ample credit with the merchants for he was the sole heir of his rich bachelor uncle, Hamish MacTavish, a canny Scot who derived his great wealth from prudent stock purchases. Further, Mr. MacTavish had suffered several heart attacks lately and his days were numbered. So Donald splurged and borrowed; his creditors knew it wouldn't be too long before they would be repaid, with handsome interest.

Donald was nice to his old uncle. Since Mr. MacTavish was a devoted gourmet (the only pleasure left to him since his physician had forbade him his beloved Scotch whiskey and his bonnie ladies), Donald had him over to his apartment for dinner once a week, trying each time to surprise him with some new epicurean delight.

One day at school during our usual Spartan lunch, sandwiched in between anatomical dissection and renal physiology, guaranteed to keep our appetites to a minimum, Donald brought up the subject of *haute cuisine*. "Say, boys, do you have any suggestions for some fancy gourmet dishes I can serve Uncle Hamish? I've just about run out of recipes I can prepare for the old man. I've gone through the whole cookbook from 'Angels on Horseback' to 'Zampino,' and I'm stuck!"

As my mind turned toward the food we ate at home, I had a sparkling inspiration. "Donald, I've got just the dish for your uncle, and one he's never had before—my grandmother's homemade matzo balls in chicken soup. She's famous for it all over Montreal. She uses a secret recipe she guards jealously. What a flavor! And those matzo balls—they're so light, they float! I almost expect them to take off from the soup dish like a falcon. I'm sure she'd prepare a potful if I ask her to."

Donald knew what chicken soup was, for he was a good cook himself, but it took some time to explain what matzo balls were. He quizzed me at some length about them. I sympathized with him: supposing the tables were reversed and he had to explain to an Orthodox Jew what a haggis was!

"Great!" exclaimed Donald, "let's plan a gourmet dinner for my uncle next week. Ask your grandmother if she'd graciously consent to prepare her famous dish of matzo ball soup and I'll plan the rest of the meal around this *piece de resistance*. And, of course, she must join us for dinner!"

Bobeh Leah agreed; how could she refuse me, her oldest grandson, and a medical student to boot! On the appointed evening Bill and I escorted my grandmother, she hugging her large pot of matzo-ball soup as if it were Fort Knox. "Bobeh," I said, "let me carry it, it's heavy."

"Listen, my grandson, I'm not that old yet that I can't carry my own cooking; besides, I don't want you mashing up the matzo balls and ruining my reputation as a cook."

Mr. MacTavish was already at Donald's apartment. After introductions, my grandmother carried the pot into the kitchen. "Donald," she said, "show me where the soup dishes are and I'll heat up my soup and serve you all at the table."

"No, no! Mrs. Starkman," Donald protested, "you're my guest. Please go into the living room and socialize with my uncle and the boys. I'll take care of everything myself."

"In that case, Donald," Bobeh replied with a professional air, "heat slowly to a simmer; put three matzo balls in each dish, and then only half fill the dish with the soup. Do it right, Donald, I don't want your uncle to be disappointed."

"Come on Bobeh," I said, "We don't want Mr. Mac Tavish to be lonely, let's go talk to him."

After a while we sat down to dinner. With a flourish Donald set before each of us a dish of my grandmother's gourmet delicacy—matzo balls in chicken soup. Mr. MacTavish was entranced. "I've never tasted anything like this, and what a flavor! Did you invent this recipe yourself, Mrs. Starkman?"

"I changed it a little, a dash of this, a sprinkle of that, but basically it's been in the family for 4000 years; during the Exodus one of our ancestors found she could do wonders with some crumbled unleavened bread, a little parsley, and an Egyptian chicken."

But conversation soon ceased as we applied ourselves assiduously to the task at hand. Suddenly, Mr. MacTavish turned a peculiar color, clutched the left side of the chest, grimaced intensely with pain, and fell forward into his soup dish. Bill, who got to him first, felt for his pulse, but shook his head and said, "I can't get his pulse, he may be dead. I'll call the ambulance, and Donald, you call Dr. Gordon Fraser who lives in the next block."

There was much commotion. My grandmother was crying. "Such a nice man," she said softly. Donald ran for the doctor, Bill was on the phone, and I was attempting resuscitation. Finally we got Mr. MacTavish to the hospital.

Later that night we heard the official report: Mr. MacTavish had died of a heart attack." Don't be too distressed, Donald," said Dr. Fraser, "after all, your uncle had had three heart attacks in as many years and it was merely a matter of time before another would carry him off." Donald seemed inconsolable.

Next day, my grandmother phoned me. "What a tragedy! I really liked that Mr. MacTavish—may his soul rest in peace. But tell me, my young doctor, did they find the poison yet?"

"Bobeh!" I exclaimed, "what are you talking about! Mr. MacTavish died of a heart attack; you knew he had heart trouble. Besides, that's the official verdict of the hospital physicians and the medical examiner!"

"So what's a 'medical examiner'?" she asked. I explained he was a special kind of doctor, a pathologist.

"So tell me, doctor"—whenever she called me "doctor" I knew it betokened some irony—"so tell me, doctor, pathologists don't make mistakes?"

"All right, Bobenu" I said, using an affectionate diminutive for "grandmother." Wishing to end on a less argumentative note I said with an attempt at humor, "You've been watching too many Charlie Chan movies. Now have a good night's sleep."

"We'll see," she said. I detected a tinge of sly smugness in her voice. "Goodnight, my grandson; don't stay up too late studying, it's not good for your health."

A few days later I got a call from the medical examiner, Dr. Henri Lafleur. Through our amateur sleuthing Bill and I had developed good rapport with him. "Dubie, your grandmother was right after all; I hope you turn out to be as good a diagnostician as she is. Mr. MacTavish died of digitoxin poisoning. We used the old reliable test, the heart of a mouse fetus; the suspect material caused the fetal heart to stop dead in systole. And you know how fatal a large dose of digitoxin can be to an already diseased heart, such as Mr. MacTavish had.

Agitated, puzzled, sheepish, I called my grandmother immediately. "Bobeh, you were right, Mr. MacTavish did die of poisoning, but what made you suspicious?"

Bobeh Leah answered smugly. "The matzo balls floated. Those in Mr. MacTavish's soup were floating; in all the other soup dishes they were resting comfortably on the bottom of the dish."

"But Bobeh!" I expostulated, "All Matzo balls float! I'm always bragging about how light your matzo balls are, how they float!"

"That's a legend perpetuated by all Jewish grandsons." Then mimicking me, "Oh, does my grandmother make the best matzo balls, they're so light they float like little balloons! So tell me, doctor, have you ever seen my matzo balls float?"

I thought awhile, "No, now that I think back; but then the soup dish was always only half filled with soup and I couldn't tell whether they floated or not."

"Aha!" said my grandmother," that's what we Bobehs do all the time. If you grandsons always insist that your Bobehs' matzo balls float, we Bobehs use this little trick so you can keep on believing it. Actually, my little doctor, all self-respecting matzo balls rest nicely on the bottom of the dish.

"Well," she continued,"when I saw the matzo balls in Mr. MacTavish's soup float, I knew something wasn't kosher. So when everyone was busy with the doctor and the ambulance I took a small sample of chicken soup and matzo balls from Mr. MacTavish's dish and put it in a jar, which I quietly slipped into my purse. So maybe I learned something from the Charlie Chan movies after all!"

Bobeh continued, "The next day I went to see the chief of detectives, Mr. Arsene Lupin—you remember him, you and Bill Osler had some dealings with him before. When he heard I was your grandmother, he saw me at once, but he looked a little funny. *'Bonjour, Madame,'* he said politely, pulling out a chair for me, 'how may I help you?' Oh these Frenchmen, always so polite, they sure know how to treat a lady. I answered, 'Mr. Lupin, we can help each other. It's about the sad death of Mr. Hamish MacTavish.' I then pulled out the little bottle from my purse and, putting it on his desk said, 'I think Mr. MacTavish was poisoned. I took this sample from his soup dish. Please examine this sample of chicken soup and matzo balls for poison.'

"'Madame, I know what is poison, I know what is chicken soup, but, forgive me, what are matzo balls?'

"'Mr. Detective, it's too complicated to explain matzo balls, but *s'il vous plait* examine this for poison.' I felt if I used a few French words he'd listen better.

"'But I have not the need,' said Mr. Lupin, 'my pathologist tells me Mr. MacTavish died of a heart attack.'

"'Mr. Lupin, *excusez-moi,* I should not contradict a famous detective like you, named after your great ancestor, but you
do have the need to examine this sample for poison.'

"'First, Mr. MacTavish's matzo balls floated while all the
others sank. I was supposed to supply that delicacy for the
evening, and I know my matzo balls never float; they sink. So
somebody was playing hanky-panky with the matzo balls in
Mr. MacTavish's soup. Second, I have a fine reputation as a
cook here in Montreal, and if the word gets out that Mr. Mac-
Tavish died eating my matzo ball soup, my reputation will be
as dead as the poor chicken who gave her life to supply the
soup. Besides, I'm a widow and I'm keeping steady company
with a certain gentleman; so you think he'll propose when he
hears I'm a Lucrezia Borgia instead of a respectable Bobeh?'

"And then I played my trump card. Looking at him soul-
fully I said, 'Mr. Lupin, I'm sure you must have a nice French
grandmother who makes the best French-Canadian pea soup in
Montreal—you call it *Soupe aux Quatorze Affaires*—delicious!'

"*Mais oui,*' he said dreamily, '*ma chere grand'mere* does
indeed make the best pea soup in Montreal.'

"'Very well, if your *grand'mere* comes to you one day,
pulls a bottle out her purse, and says to you "*Mon cher garçon,*
one must examine this pea soup for poison"—what would you
do?'

"'*Eh bien, madame,* I would do it, of course.'

"'*Eh bien, monsieur,*' I echoed, 'just this once imagine
I'm your Bobeh—excuse me, *grand'mere,* and please examine
my soup for poison.'"

Mr. Lupin gave in graciously, and with resignation, for he
knew he couldn't cope with my grandmother's convoluted tal-
mudic thinking. "Very well, madame, I shall have the sample
examined by the toxicologist."

"And that is how the sample got to the medical examiner,
my young doctor."

"So who did it, Bobeh?" I asked

"Use your common sense, your logic," she answered.
"People say you shouldn't kill the goose that lays the golden

eggs, but what if the goose is an old gander (may his sould rest in peace!) and he's worth more gold dead than alive? If I were a gambling person I'd put my money on that no-goodnik Donald. He couldn't wait till his poor uncle died, he had to give him a nudge. He made his own pot of chicken soup and matzo balls just for his uncle, but he added a little nasty ingredient I never include in my own recipe, God forbid!

"But why did he make the matzo balls float? And how did he *do* it, especially since you say all matzo balls *sink?*

"He made them float because you had him brainwashed!" Here she mimicked me again, "Oh Donald, does my grandmother make the best matzo balls, they're so light—they float!' And how? That's an old trick: if you used soda water instead of plain water in the recipe the bubbles of gas get trapped inside the dough—and the matzo balls float! Every time my neighbor, Mrs. Goldberg, wants to impress a new boy friend she uses the old soda-water trick. She should be ashamed, no honest cook would do it." She sniffed contemptuously.

"Well, Bobeh, I said humbled, "I must admire how you solved the case so cleverly when everyone else was fooled, including Detective Lupin, the medical examiner, and even my brilliant friend Guillaume."

"My little doctor," she said affectionately, "it should be a lesson for your future career in medicine. I've watched some of you young doctors, swinging your stethoscopes, running for laboratory tests, using your fancy machinery; you have to have a chest X-ray and a Bagel test before you can diagnose a cold."

"Bobeh, please," I corrected her, "that's not a Bagel test, it's a Nagel test."

"So now you're going to flunk your old Bobeh in spelling? No, my dear grandson, I'm trying to tell you what really counts is observation, logic, and common sense. The rest is just fency-shmensy, like a bridal veil, but what's underneath? You think I didn't read Arthur Cohen Doyle and his stories about his famous detective Shylock Holmes? Did you know Cohen Doyle was a medical doctor, just like you?

"And who was the model for Shylock Holmes? His teacher, Dr. Bell, who was famous for his observation and deduction. And from whom did Dr. Doyle get his wonderful logical approach to the problems, who inspired him? Guess! from his Bobeh Cohen! And Shylock, did he use fancy equipment? No—a little chemistry and an old microscope—but always observation, observation, and logic, logic! I'll bet his grandfather must have drummed the Talmud into his head.

"So you see, my little student, I had no gadgets, I used only my eyes, my nose and my common sense."

"But, Bobeh, this is the first time you mention using your nose in this case!"

"Oh yes, I'm getting forgetful. When I was examining Mr. MacTavish's chicken soup I smelled it. And I smelled a big rat, because I didn't smell any Dill. I knew something wasn't kosher, because tell me, my grandson, have you ever heard of any self-respecting Jewish chicken soup that doesn't smell of Dill?"

I. N. Dubin
Perspectives in Biology and Medicine
Vol. 22, Autumn 1978, pp. 127–133

Discovery of a New Radiation Source Z-I in Taurus

Summary

A new (and only) source of celestial Z-radiation has been detected. Possible implications and indentifications are discussed.

Introduction

For several years, attempts have been made to detect a new type of radiation in space (known as Z-radiation), but because detection equipment was below the required threshold as a result of atmospheric absorption, and because the rapid accelerations of rocket flights prevented the graduate students in the nose cones from recording good data, this has not hitherto been achieved. We now report the first positive identification of celestial Z-radiation.

The United States satellite Zolo-I was launched into polar orbit on 1973 August 23 from Vandenberg Air Force Base. Although originally designed as an Earth resources satellite to monitor the frequency of Peruvian guano birds, one of the 312 lead sulfide LSD-doped wide-angle detectors was of sufficiently broadband to record a Z event (see Fig. 1). Although only one event was recorded, the statistical theory of the single-event event (SEE) has been developed enough in recent years in various branches of space research to provide much information on such occurrences. (The fact that *none* of the other 311 tandem detectors, nor anyone else, observed the source clearly confirms its SEE nature.)

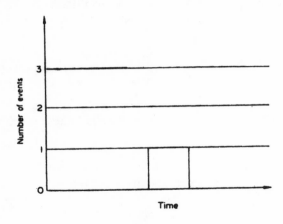

Fig. 1

Position

Since the detectors lacked time resolution of better than 0.1 s, one can only set an upper limit on the pulse duration. However, by assuming that half the pulse occurred in the first 0.05 s and half in the last 0.05 s, and knowing the light travel-time across the satellite's orbit is 0.1 s, one can locate the SEE in the field to within ±5°. This places the SEE in one of two intersections of great circles, but the ambiguity can be removed since only one position falls over Cambridge, Massachusetts. (The other position is pure Taurus.)

Spectrum

The time-pulse structure was approximated by a delta function, the Fourier transform of which revealed the SEE spectra (*see* Fig. 2). Although the spectrum is very nonthermal, the fact that there is no low-frequency cutoff implies that synchrotron self-absorption is negligible. Confirmation of this by radio observations at lower frequencies would be useful. As a check on the consistency of the data, the signal was divided in half symmetrically and folded on itself (*see* Fig. 3). The resulting fit proved to be excellent. No spectral absorption lines have been identified as yet, but this may reflect a sufficiently strong gravitational potential that would smear out the absorption lines below our detection limits.

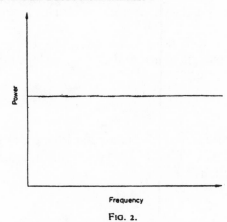

FIG. 2.

Possible Optical Identifications

No optical objects brighter than 18th mag lie within the 5° error box on the Palomar Sky Survey. An airplane carrying a lighted sign passed just north of the error box as seen from Agassiz station on 1973 October 2, but its measured blue excess (B − V = ±0.02) is not sufficiently anomalous to warrant positive identification.

Possible Chance Occurrence

Although only one SEE was seen, the possibility exists that there were really two superimposed sources. However,

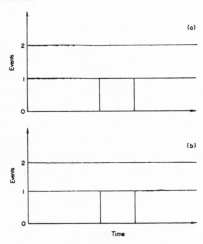

Fig. 3. Before (a) and after (b) folding the data.

the probabililty that this is the case has been independently calculated by two different authors as 10^{-143} and 0.02, respectively, so that it may be neglected.

Isotropy

The distribution of all known Z-radiation in galactic longitude is illustrated in Fig. 4. The distribution is isotropic except in the 3rd quadrant.

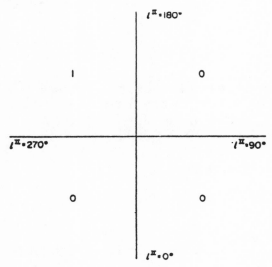

Number of Sources in the Sky

Since we observed the SEE of 0.1 s and the age of the Universe is 10^{10} yr (= 3×10^{17} s), if each galaxy goes through a Z-active phase there must be 3×10^{18} SEEs undetected now in the sky. In cosmological terms, the argument is altered only slightly by a factor of $(1 + z)^4$, where z is the fractional redshift in a Universe for which $q = +1$.

No linear or elliptical polarization was observed during the SEE, since none was looked for.

Periodicity

No evidence for periodicity has been observed.

$S^{-2/3}$ Distribution

The SEE defines its own $S^{-2/3}$ curve, on which it therefore lies exactly. Thus, it is either a very close object ($r = 1$ cm, flux = 10^{-10} fu) or very cosmologically distant ($r = 10^{81}$ cm, flux = 10^{152} fu).

Polarization

No linear or elliptical polarization was observed during the SEE, since none was looked for.

Discussion

None of the known SEEs coincides with the centers of rich galactic clusters, H II regions, highly dust-obscured stars, or radio sources. None was coincident with Weber's gravitational events. Since the lack of synchrotron self-absorption rules out a supernova, the SEE is clearly a black hole. The estimated 3×10^{18} SEEs easily provide enough mass to close the Universe.

From the data presented here it has been calculated that a detector of area corresponding to the state of Massachusetts would be needed to increase the detection frequency significantly. A proposal is therefore being submitted to NASA that the state of Massachusetts should be wired with Z-detectors every 3 cm^2 and launched into orbit to search for the remainng 3 x 10^{18} − 1 SEEs. It is expected that there will be strong governmental support for this project.

Larry S. Liebovitch
*Quarterly Journal of the Royal
Astronomical Society,* London
Vol. 15, 141–145 (1974)

How Insects Hear

*An Evolutionary Thought on the Variety of
Hearing Organs in Nocturnal Insects*

In days of old and insects bold
(Before bats were invented),
No sonar cries disturbed the skies—
Moths flew uninstrumented.

The Eocene brought mammals mean
And bats began to sing;
Their food they found by ultrasound
And chased it on the wing.

Now deafness was unsafe because
The loud high-pitched vibration
Came in advance and gave a chance
To beat echolocation.

Some found a place on wings of lace
To make an ear in haste;
Some thought it best upon the chest
And some below the waist.

Then Roeder's keys upon the breeze
Made hawk moths show their paces.
He found the ear by which they hear
In palps upon their faces.

Of all unlikely places!

J. D. Pye
Nature
Vol. 218, May 25, 1968

The Last Pterodactyl

Doyle's hero Professor Challenger believed that prehistoric animals still lived in South America. He and a party of friends undertook a perilous journey to prove this. On their return they tried to convince an unruly and skeptical audience of their findings.

Professor Challenger: "Passing then, both the photographs and the entomological collection, I come to the varied and accurate information which we bring with us upon points which have never before been elucidated. For example, upon the domestic habits of the pterodactyl—"(A voice: "Bosh," and uproar)—"I say, that upon the domestic habits of the pterodactyl we can throw a flood of light. I can exhibit to you from my portfolio a picture of that creature taken from life which would convince you—"

Dr. Illingworth: "No picture could convince us of anything."

Professor Challenger: "You would require to see the thing itself?"

Dr. Illingworth: "Undoubtedly."

Professor Challenger: "And you would accept that?"

Dr. Illingworth (laughing): "Beyond a doubt."

It was at this point that the sensation of the evening arose —a sensation so dramatic that it can never have been paralleled

in the history of scientific gatherings.   Professor Challenger
raised his hand in the air as a signal, and at once our colleague,
Mr. E. D. Malone, was observed to rise and to make his way
to the back of the platform.   An instant later he reappeared in
the company of a gigantic Negro, the two of them bearing
between them a large, square packing-case. It was evidently of
great weight, and was slowly carried forward and placed in
front of the Professor's chair.   All sound had hushed in the
audience and everyone was absorbed in the spectacle before
them.   Professor Challenger drew off the top of the case,
which formed a sliding lid. Peering down into the box he snap-
ped his fingers several times and was heard from the Press seat
to say, "Come, then, pretty, pretty!" in a coaxing voice.   An
instant later, with a scratching, rattling sound a most horrible
and loathsome creature appeared from below and perched itself
upon the side of the case. Even the unexpected fall of the Duke
of Durham into the orchestra, which occurred at this moment,
could not distract the petrified attention of the vast audience.
The face of the creature was like the wildest gargoyle that the
imagination of a mad medieval builder could have conceived.
It was malicious, horrible, with two small red eyes as bright as
points of burning coal.   Its long, savage mouth, which was
held half-open, was full of a double row of shark-like teeth.
Its shoulders were humped, and round them was draped what
appeared to be a faded gray shawl. It was the devil of our childhood in person. There was a turmoil in the audience— someone screamed, two ladies in the front row fell senseless from
their chairs, and there was a general movement upon the platform to follow their chairman into the orchestra. For a moment
there was danger of a general panic. Professor Challenger
threw up his hands to still the commotion, but the movement
alarmed the creature beside him. Its strange shawl suddenly
unfurled, spread, and fluttered as a pair of leathery wings. Its
owner grabbed at its legs, but too late to hold it. It had sprung
from the perch and was circling slowly round the Queen's Hall
with a dry, leathery flapping of its ten-foot wings, while a
putrid and insidious odor prevaded the room. The cries of the
people in the galleries, who were alarmed at the near approach
of those glowing eyes and that murderous beak, excited the
creature to a frenzy. Faster and faster it flew, beating against

the walls and chandeliers in a blind frenzy of alarm. "The window! For heaven's sake shut that window!" roared the Professor from the platform, dancing and wringing his hands in an agony of apprehension. Alas, his warning was too late! In a moment the creature, beating and bumping along the wall like a huge moth within a gas-shade, came upon the opening, squeezed its hideous bulk through it, and was gone. Professor Challenger fell back into his chair with his face buried in his hands, while the audience gave one long, deep sigh of relief as they realized that the incident was over....

One word as to the fate of the London pterodactyl. Nothing can be said to be certain upon this point. There is the evidence of two frightened women that it perched upon the roof of the Queen's Hall and remained there like a diabolical statue for some hours. The next day it came out in the evening papers that Private Miles, of the Coldstream Guards, on duty outside Marlborough House, had deserted his post without leave and was therefore courtmartialed. Private Miles' account, that he dropped his rifle and took to his heels down the Mall because on looking up he had suddenly seen the devil between him and the moon, was not accepted by the Court, and yet it may have a direct bearing upon the point at issue. The only other evidence that I can adduce is from the log of the *SS Friesland,* a Dutch-American liner, which asserts that at nine next morning, Start Point being at the time ten miles upon their starboard quarter, they were passed by something between a flying goat and a monstrous bat, which was heading at a prodigious pace south and west. If its homing instinct led it upon the right line, there can be no doubt that somewhere out in the wastes of the Atlantic the last European pterodactyl found its end.

Arthur Conan Doyle
The Lost World
Chapter XVI
Hodder & Stoughton, 1912

Chemical Engineering in the Kitchen

The "art" of cooking has its traditions steeped in history and, like the arts in general, can trace its origins to prehistoric humans. The discipline of chemical engineering, which at times also takes on the semblance of an art (perhaps more often than we might care to admit), is more of a 20th century phenomenon. Even so, some of the principles embodied in its curriculum can be traced to earlier centuries.

Perhaps this article will infuriate both chemical engineers and expert cooks by suggesting comparisons between the two disciplines. But I apologize to neither group.

The kitchen and its immedite surroundings are a veritable laboratory, combining both a bench-scale and pilot-plant atmosphere. The astute chemical engineer will quickly learn that certain basics of his or her craft can easily be applied to the art of cooking—with a surprising degree of success. After a while, one likewise discovers that certain principles and techniques of cooking can be readily adapted to those of chemical engineering design.

Engineers intimately familiar with *Transport Phenomena* by Bird, Stewart, and Lightfoot might recall some special problems dealing specifically with the kitchen. One in particular that I recall was the mathematical relationship between the mass of a turkey and its cooking time. This was by no means the intitial application of chemical engineering to the culinary art, but it did inspire others to proceed along the same lines.

A friend of mine earned his master's degree working on a similar mathematical model for baked potatoes. Unfortuntely, during the course of experimentation, he inadvertently gained thirty-five pounds. But people have surely suffered graver consequences in the pursuit of science.

I must admit, the research for most of the examples to follow are a result of my own experiences puttering around the kitchen, both as a hobby and as a necessity (I'm single).

It wasn't, however, until I became involved in the classical chemical engineering problems that I began to recognize the association between the seemingly different "sciences."

Phase Separation and Soupmaking

As a youth, I realized that soups made from meat stocks always tasted better the second day. It wasn't until later that I realized the reason for this was that, after being placed in the refrigerator overnight, the broth and fat globules separated into two very distinct layers. The fat, of lesser specific gravity, coagulating on top, could then be easily scraped off before serving. The cooling of the mixture in the refrigerator caused the soup to gel, forming the two layers. Even if the fat is continually skimmed off the surface during the course of cooking the soup, it is impossible to separate all of the fats and oils from the broth. However, the "fractional crystallization" technique employed after the soup is placed in the refrigerator makes this phase separation quite easy.

How unfortunate those chemical engineers were trying so desperately to separate *ortho-* and *para*-xylenes with a huge and complicated array of distillation equipment because they had never made chicken soup. Otherwise, they surely would have attempted fractional crystallization sooner!

Spaghetti and Surfactants

Those of us who have frolicked in our first attempts at spaghetti will probably remember the persistent problems of foaming while cooking the pasta (not unlike the problems of foaming during certain polymerizations). Experience demonstrated (to both the chef and the scientist) that the addition of a tablespoon of some type of vegetable oil kept the problem nicely under control. Thus, we could tend to the more important aspects—such as the spaghetti sauce. This is one surfactant even the chemical process industries can't claim to have developed. But just as effectively as their surfactants help control foaming in polymer reactors, so does one tablespoon of vegetable oil in a pot of boiling pasta.

A little deeper investigation also reveals that it is the protein in the pasta that causes the foaming. In point of fact, a friend from Eli Lilly informs me, solutions of proteins foam vigorously when stirred and will polymerize at elevated temperatures.

Culinary Unit Operations

Close observation in the kitchen will identify many applications for the varied unit operations and techniques used by our profession. Are not the oven and pressure cooker types of reactors: the Waring Blendor, an agitator, slicer, dicer, chopper, and mixer; the colander, a simple dewatering device? These analogies, I suspect, anger even the amateur chef no end. The mere mention of scientific gadgetry as applied to the craft sends him or her into a frenzy. Yet the recipes (formulas?) are kept vague and confusing.

I feel certain the nuances of cooking can, in many cases, be explained in absolute terms as opposed to "a pinch of this" or "a sprinkle of that." Measured viscosities would do a great deal to better define the making of the multitude of sauces and gravies. Certainly one could more easily relate to a "10 centipoise" bearnaise or a "500 centipoise" hollandaise after a brief explanation of the concept of viscosity. This would much improve one's ability to prepare sauce of the proper consistency each time. I have often thought that, besides introducing more-explicit terms such as viscosity, a well-placed erlenmeyer flask or graduated cylinder would also greatly improve quality control in the kitchen.

Kitchen Science

As one gets better acquainted with the kitchen and its appropriate techniques, one notices the principles of chemical engineering in particular, and science in general, being demonstrated beautifully and subtly. The exhilarating Tyndall effect is quickly identified as light beams are dispersed by a glass of Jello (Jello is a perfect colloid).

One of the more interesting scientific principles is demonstrated when a carbonated beverage is placed in the freezer and later taken out and opened. If the temperature is just right, the solution will remain a liquid until the cap is removed—then in a moment the contents will begin to rapidly freeze, from top to bottom. The Joule-Thomson effect may be playing an integral part in this phenomenon.

Even the precise laws of thermodynamics are demonstrated daily in the kitchen. These laws, which have gained universal application transcending the traditional sciences, are now applied to disciplines such as economics. But nowhere do they manifest themselves more significantly than in the culinary field.

What better way to understand the abstract concept of entropy, as introduced in the Second Law, than to apply it to the instance where a filet mignon is tragically burnt to a crisp. Surely it can never return to an edible state (or, more precisely, uncook itself). Perhaps I should say that the irreversibility of such a process is dictated by the concept of entropy. Even more accurately, I should have to say that the probability of such an "uncooking" process is exceedingly small (and you thought entropy had no practical application). The next time your spouse has rendered a meal unpalatable from overcooking, don't blame her (him) for not being able to restore it. Blame J. Willard Gibbs.

What I have discussed in this article was not meant to be facetious. The facts presented contain principles of chemical engineering as applied to the techniques of what we might call "culinary science," and also principles of "culinary science" as applied to what we could call chemical engineering. Perhaps I did annoy both gourmet and chemical engineeer, but the cold facts of science are sometimes harsh and cruel.

Benjamin A. Horwitz
Chemical Engineering
August 9, 1982; p. 65

A Peerless Contraceptive

. . . The details of the synthesis and structure of the drug are protected by patent considerations, but we can disclose the pertinent essentials. The synthesis begins with the reduction of nitrobenzene in alkaline solution to bring about the necessary coupling. Because of the reducing agent employed, the product is not azoxybenzene, azobenzene, or hydrazobenzene as might be expected, but a new compound whose molecular configuration can be revealed to the extent shown in Fig. 1.

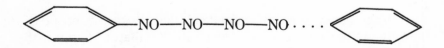

Fig. 1. Chemical configuration of the reactive portions of the Armpitin molecule Armpitin-4.

. . .the layman can quickly identify the —NO— groups and relate these to the obvious contraceptive connotation. What is even more remarkable, the synthesis can be readily controlled so that the number of —NO— groups may be regulated from one to infinity. This in itself would not be so extraordinary were it not for the fact that both extensive animal and clinical trials have clearly demonstrated that for each additional —NO— group in the backbone of the structure there is a corresponding one-day effectiveness of the drug! Thus, if there are four —NO— groups (as illustrated), the drug is absolutely effective for four days and no longer. This has limitless commercial and therapeutic possibilities that we will exploit fully.

We have chosen to designate the drug "Armpitin" since in our clinical experience its effectiveness is most pronounced when it is applied to the female axillary regions. The drug can be readily incorporated into a roll-on, spray, cream, or any other convenient form of applicator currently used by commercial deodorant manufacturers.

Julius S. Greenstein
Chemtech
April 1979; pp. 217-221
from *Canadian Medical Association Journal Vol. 93, December 1965*

Properties and Composition of Lunar Materials
Earth Analogies

The sound velocity data for the lunar rocks were compared to numerous terrestrial rock types and were found to deviate widely from them. A group of terrestrial materials were found that have velocities comparable to those of the lunar rocks, but they do obey velocity–density relations proposed for earth rocks.

Certain data from Apollo 11 and Apollo 12 missions present some difficulties in that they require explanations for the signals received by the lunar seismograph as a result of the impact of the lunar module (LEM) on the lunar surface. In particular, the observed signal does not resemble one due to an impulsive source, but exhibits a generally slow buildup of energy with time.

In spite of the appearance of the returned lunar samples, the lunar seismic signal continued to ring for a remarkably long time—a characteristic of very high Q material. The lunar rocks, when studied in the laboratory, exhibited a low Q. Perhaps most startling of all, however, was the very low sound velocity indicated for the outer lunar layer deduced from the LEM impact signal. The data obtained on the lunar rocks and fines agree well with the results of the Apollo 12 seismic experiment. These rock velocities are startlingly low. The measured velocities on a vesicular medium grained, igneous rock (10017) having a bulk density of 3.2 g/cm³ were $v_p = 1.84$ and $v_s = 1.05$ km/sec. The results for a microbreccia (10046) with bulk density of 2.2 g/cm³ were $v_p = 1.25$ and $v_s = 0.74$ km/sec for the compressional (v_p) and shear (v_s) velocities.

It was of some interest to consider the behavior of these lunar rocks in terms of the expected behavior based on measurements of earth materials. Birch first proposed a simple linear relation between compressional velocity and density for rocks.

These materials are summarized in Table 1 (not shown here), where, for emphasis, common rock types found on earth are listed for comparison. The materials studied were chosen so as to represent a broad geographic distribution in order to preclude any bias that might be introduced by regional sampling. It is seen that these materials exhibit compressional velocities that are in consonance with those measured for the lunar rocks—which leads us to suspect that perhaps old hypotheses are best, after all, and should not be lightly discarded.

A comparison of these low velocity materials with the predictions of Birch and Anderson is shown in Fig. 1. It is at once apparent that these materials do yield values of velocity that are predicted by these relations for their densities. Thus the curve of Birch for the rock types diabase, gabbros, and eclogites fit the cheeses surprisingly well. This apparent inconsistency, in that the cheeses do obey these relations by having a velocity appropriate to their density, in contrast to the lunar rocks with which they compare so well, may readily be accounted for when one considers how much better aged the lunar materials are.

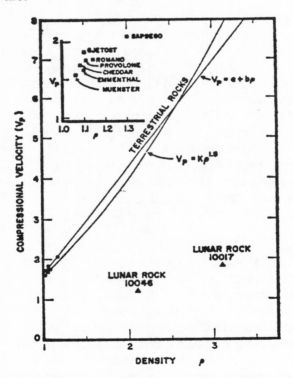

Fig. 1. Comparison of the velocity of sound for rocks with that of earth materials.

Edward Schreiber and **Orson L. Anderson**
Science, Vol. 168, 1579 (1970)

Dyson Civilizations

A search could now be made for the so-called "Dyson civilizations," according to an article in a recent issue of the *Astrophysical Review*. Dyson civilizations are intelligent beings in another solar system who have knocked their planets to pieces and used the materials to build a hollow ball around their sun. This intriguing idea was first put forward in 1960 by Dr. Freeman Dyson, the theoretical physicist who works at the Institute for Advanced study at Princeton.

One reason anyone should want to carry out such an operation would be the pressure of an expanding population. A hollow sphere would certainly represent a great economy in the use of the heat and other energy emitted by the central star, most of which is wastefully radiated into space under the old-fashioned method.

But such shells would still radiate a great deal of heat, and this emission might be detected by looking for radiation at wavelengths between 8 and 13 microns with a powerful telescope, such as the infrared telescope at Mount Wilson. There would be a risk of confusing them with the newly discovered "cool stars" about which I wrote last January, but other methods could be employed to establish differences. Having located a likely object, one would then listen for radio signals from it.

The suggestion that available equipment is sufficiently sensitive to detect the infrared radiation from a huge hollow shell comes from Dr. Carl Sagan, then at the Harvard College Observatory. He published a paper that outlined a search pattern, designed with Russell G. Walker, a graduate student at the Air Force Cambridge Research laboratories, Bedford, MA. They considered that we could survey all infrared systems within 1000 light years of the sun, and would have "a good chance" of finding "at least several advanced technical civilizations or equivalent natural objects"—whatever these might be.

After 20 years, Dr. Sagan still actively pursues concrete evidence of such extraterrestrial intelligence (ETI). And though the probability statistics seem promising; though he has helped the US Government to support monitoring the heavens for messages from ETIs; though he has even convinced NASA to send trinkets identifying the origin of our space probes and the nature of the people who have sent them, there has as yet been no response. Perhaps all ETIs live on the inner surface of Dyson shells and ever focus their thought inward toward the central star....

Science Journal, vol. 2
London, November, 1966

Friday Experiments

Everyone knows that on the island with Robinson Crusoe was his man Friday. Not generally known was the fact that Friday was a scientist who performed many elaborate experiments, especially with gases. Now these facts have come to light because the noted anthropologist Sir Humphrey Harvey has recently discovered the notebook in which Friday recorded his research. Harvey had great difficulty in deciding whether Friday's experiments were worthwhile or not because of the units in which Friday recorded his data. For example, Friday measured pressure by collecting bird droppings on a barrel stave affixed to his apparatus. Consequently, all of this pressure readings are in bird droppings per barrel stave (Bd/Bs) units. By carefully reproducing some of Friday's experiments (with pigeons), Harvey determined that 15 torr = 1.70 Bd/Bs when the bird dropping was confined to an area on the end of the barrel stave measuring 13 x 16 mm. Similarly, Friday used old goat bladders to contain his gases, and consequently all volume readings are in units of old goat bladders (Ogb). By averaging the value obtained from ten old goats, Harvey found that 1.00 Ogb = 265 Bombay Gin bottles. After an interval of two months, Harvey then determined that 1.00 Ogb (Bombay Gin bottle) = 0.85 L.

Friday was aware of Avogadro's law, though not by name. Harvey is convinced Friday rediscovered it, as well as Boyle's and Charles' Laws. However, Friday measured numbers of gaseous particles by a unique, but reproducible, method. He would carefully introduce a leak into the old goat bladder, squeeze it forcefully and note the high-pitched screech as the gases were exhaled. This procedure invariably attracted chipmunks, which abound on the island. Since the number of chipmunks that rushed to the opening of the cave (which was the location of Friday's laboratory) was directly proportional to the sound frequency and this was a function of the numbers of molecules, Friday's unit for number of molecules was measured in number of chipmunks (Chip). Harvey determined that 12 Chip = 1.00 mole.

Friday measured temperature as follows: he would suspend an old goat bladder in the shade and measure its volume by an unclear method (that portion of the manuscript was eaten away, apparently by chipmunks). Then, suspending the old goat bladder in a vat of gently boiling papaya slurry, he found the volume was increased to 7/5 its original value. By dividing this difference in temperature into degrees papaya slurry ($°Ps$), Friday concocted a temperature scale. Harvey reproduced these experiments and found that a tempature range of $11°Ps$ corresponds to a temperature range of $23°C$, and the freezing point of water is $-14°Ps$ on Friday's scale. Thus, absolute zero is $-145°Ps$. Harvey ingeniously devised a new unit called absolute degrees papaya slurry ($°Ps–Abs$) that starts at absolute zero = $0°$ and, naturally the freezing point of water = $131°Ps–Ab$.

Apparently, Friday's most detailed and discriminating experiments were concerned with a substance which he called "banana gas." A facsimile of one of Friday's data sheets is given in the table.

Further work on banana gas was performed by Friday in an attempt to determine some of its physical properties. In one of his experiments he found that a 2-Chip sample of banana

Some Data from Friday's Notebook

Bird droppings/ Barrel stave	Degrees papaya slurry	Chipmunks	Old goat bladders
23	12.2	3	1
28	17.6	7	2
40	15.5	5	1
38	42.0	4	1
23	10.3	12	4
26	29.7	6	2

gas contained in one old goat bladder and maintained at a temperature of 13.5°Ps exerted a pressure of 16 Bd/Bs. Friday weighed the empty old goat bladder prior to the experiment and found that it weighed 97.35 Bd. It weighed 98.57 Bd when it contained the 2-Chip of banana gas.

Sir Harvey has appealed to his chemist friends in the colonies to supply answers to the following questions:

(1) Is banana gas ideal?
(2) What is its molecular weight?
(3) What is its chemical formula?

He seems doggedly convinced that enough data was collected by Friday for such an analysis. He even presumes (in a leap of the imagination) that Friday had performed such an analysis. However, this is mere conjecture on his part since those pages in the notebook are missing.

Jack E. Bissey
Journal of Chemical Education
Vol. 46, p. 497, August, 1969

The Business Outline of Astronomy

The world or universe in which we do our business consists of an infinite number, perhaps a hundred billion, perhaps not, of blazing stars accompanied by comets, dark planets, asteroids, asterisks, meteors, meteorites, and dust clouds whirling in vast circles in all directions and at all velocities. How many of these bodies are habitable and fit for business we do not know.

The light emitted from these stars comes from distances so vast that most of it is not here yet. But owing to the great distance involved, the light from the stars is of no commercial value. One has only to stand and look up at the sky on a clear starlight night to realize that the stars are of no use.

Practically all our efficient light, heat, and power comes from the sun. Small though the sun is, it gives out an intense heat. The businessperson may form some idea of its intensity by imagining the entire lighting system of any two great American cities grouped into a single bulb; it would be but little superior to the sun.

The earth revolves around the sun and at the same time revolves on its own axis, the period of its revolution and the rising and setting of the sun being regulated at Washington, DC. Some years ago the United States government decided to make time uniform and adopted the system of standard time; an agitation is now on foot—in Tennessee—for the lengthening of the year.

The moon, situated quite close to the earth but of no value, revolves around the earth and can be distinctly seen on a clear night outside the city limits. During a temporary breakdown of the lighting plant in New York city a few years ago, the moon was quite plainly seen moving past the tower of the Metropolitan Life building. It cleared the Flatiron building by a narrow margin. Those who saw it reported it as somewhat round, but not well shaped, and emitting an inferior light that showed it probably to be out of order.

The planets, like the earth, move around the sun. Some of them are so far away as to be of no consequence and, like the stars, may be dismissed. But one or two are so close to the earth that they may turn out to be fit for business. The planet

Mars is of especial interest inasmuch as its surface shows traces of what are evidently canals that come together at junction points where there must be hotels. It has been frequently proposed to interest enough capital to signal to Mars, and it is ingeniously suggested that the signals should be *sent in six languages*.

Stephen Leacock
Winnowed Wisdom
Dodd, Mead, New York, 1926, p. 23

Mama-thematics

1. Mother to Pascal: "When you got involved with those gamblers and into that triangle, I was afraid that my Blaise would become blasé."
2. Mother Einstein to Albert: "You see, I always told you to spend less time playing that violin and to think more about relatives."
3. Mother Khayyam to son Omar: "Your algebra and astronomy may have been helped by that jug of wine, but they didn't bring in much bread."
4. Mother to Archimedes: "I do declare, you must be in your second childhood, playing in that sand-box all day!"
5. Mama-san to son Y. Yoshino: "In this computer age, I'm glad that you are loyal to your abacus. You can count on it."
6. Napier's mother to her son: "What's all this talk I hear about your bones and legs? Anybody'd think you had a wooden leg."
7. Mama snake to serpent in the Garden of Eden: "You are a calculating reptile, but you're no adder."

Charles W. Trigg
Journal of Recreational Mathematics
Vol. 13, p. 49 (1980–1981)

Concerning Index of Refraction and Density

Almost any general physics textbook one examines will either state or imply that when a ray of light passes from one medium obliquely into another medium of a different density, the angle made by the ray with the normal is less in the more dense medium. Such a statement is, at best, very misleading.

Two illustrations that indicate the actual situation are well known to most instructors. The first is that of the oil-immersion microscope objective, where glass having an index of about 1.50 and a density of 2.3 to 2.4 g/cm³ is matched approximately by oil of cedarwood ($n = 1.51$) or Nujol ($n = 1.47$), although their densities are much smaller than that of glass.

As another illustration, we have the test to distinguish Pyrex glass from soft glass by immersion in a solution of 41% benzene and 59% carbon tetrachloride (density 1.30 g/cm³) or in trichloroethylene (density 1.46 g/cm³), both of which match the index value of Pyrex (1.48) although the density of Pyrex is 3.54 g/cm³.

In view of these experimental facts, it seemed desirable to look into the matter more thoroughly, by consulting tables for values of index of refraction and density at a common temperature. Such values were obtained from the 34th edition of the *Handbook of Chemistry and Physics,* using values for 20°C and D light indices chiefly. A graph was then prepared in the hope of finding an explicit answer to the question, does index of refraction vary directly with density? The answer is found in Fig. 1.

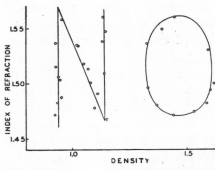

FIG. 1. Does index of refraction vary directly with density?

E. Scott Barr
American Journal of Physics
Vol. 23, December, 1955 p. 623

New Food, Fiber, and Fuel Crop

With the rapidly growing imbalance in the world food-population equation, there is an urgent need, as many social scientists have noticed, for a highly nutritious, high-yielding crop plant with wide adaptation, horizontal resistance to pests and diseases, tolerance of environmental stresses and weed competition, not requiring fertilizers and other inputs beyond the reach of Third World farmers, and preferably suitable for total harvest and biomass conversion to liquid fuel. We report here the creation of such a crop.

Although 10,000 years of traditional plant breeding have failed to achieve this goal, we believed that by deploying the potent armory of genetic engineering techniques to the full, this exciting objective could be attained within a few years. Such has proved to be the case, and this preliminary note records our general approach to the problem.

The crucial strategic decision in our experimental attack was to begin our molecular plant breeding program, not with a modern, high-yielding cereal variety, but with a wild Australian grass that has evolved under low water and nutrient conditions and therefore has no undesirable dependence on irrigation and fertilizers. The only previously known use of the species was that the seeds were gathered for fertility rites by a now almost extinct aboriginal tribe.

On this unsophisticated receptor we then brought to bear the full power of genetic engineering, recombining genes from a wide range of sources. The ability to fix atmospheric nitrogen symbiotically was incorporated, of course, in this case from the New Guinea rain forest tree *Parasponia rugosa*, and the capacity to extract phosphorus from deficient soils was recruited from a mycorrhizal fungus. Salt tolerance came from a halophytic grass, by the incorporation of salt glands. Exceptional productivity derived from a high-specific-activity form of rubisco, the key photosynthetic enzyme, found in *Sainpaulia ionantha*. High nutritional quality comes from genes for egg protein in *Hulso huso,* the giant belugo sturgeon, which could deal a severe blow to a major Russian export. Of even greater moment, however, is the incorporation of genes for an estrogenic steroid and for penicillin. The former, secured from a somaclonal variant of the Mexican yam, will ensure that the improvement in world food supply made possible by this crop will not be cancelled by equivalent population growth: the latter, derived from Fleming's original 007 strain of Penicillium, will likewise ensure the health of people under even the most primitive conditions.

Construction of the comprehensive vector, based on the T1 plasmid, is described in beguiling but incomplete detail elsewhere (O'Type, 1983). So profoundly have we reengineered this plant that we consider it deserves specific, if not higher, taxonomic rank. We have therefore named it in honor of the founders of the modern period of biological creation (*Pseudocerealis boyer-coheni*). In passing we note, for the benefit of the Nobel Prize Committee, that our new crop not only promises to solve the world food, health, and population problems, but is also the first organism to combine plant, microbial, and animal genes. It retains one recalcitrant disadvantage, which may however prove of value in reinforcing the effects of the steroid birth control component. Despite all efforts, the seed when ingested caused a profound and long-lasting halitosis, apparently associated with an analog of armpitin (Greenstein, 1965).

"Where now?" readers might well ask. That will depend, for our research group as for others, on the induced enthusiasms and perceived weaknesses of the various bodies funding such dynamic research. Meanwhile, in the pregnant words of the originators of all this activity (Watson and Crick, 1952), "It has not escaped our notice that" the results described here may be of considerable commercial interest. We acknowledge the Trappist support of this work, under grant GE 33 TATA, by Supergene of Silicon Valley.

References

O'Type, G., 1983. Construction of a universal vector for crop improvement. *Acta Retracta,* in press.

Greenstein, J. S., 1965. Studies on a new peerless contraceptive agent: a preliminary final report. *Can. Med. Assoc. J.* **93**, 1351–1355.

Watson, J. D., and F. H. C. Crick, 1952. A structure for deoxyribose nucleic acid. *Nature* **171**: 737–738.

BioScience
Vol. 34, April 1984, p. 213

This paper was written in the hope that it might induce some humility in our molecular plant breeding whiz kids. It dented them, but only temporarily. The authorship had, unfortunately, to be anonymous.

Gene O'Type, Phene O'Type, and Kary O'Type
Institute for Unconstrained Plant Breeding
Darwinia, 2999 Australia

===

Novel Device Described in
UK Patent Application

From M. R. Johnson of Midland, Michigan, has come a copy of a United Kingdom patent application (GB 2,009,924 A) on an "apparatus for identifying atomic numbers of elements." The applicant, one T. B. Orton, "has propounded the theory that all individual elements emit radiations by which they may be identified. These radiations will for the purposes of description be termed Ortonarays." The rather remarkable capabilities of Orton's device depend primarily on its ability to detect and measure these Ortonarays.

The physical nature of the apparatus is shown clearly by the drawings in the patent application, and a detailed description is beyond our scope here. Two points might be mentioned however. The instrument is equipped with a probe that can be placed against or inserted into unknowns when convenient. And to work properly, it must be oriented correctly relative to magnetic north and so is equipped with a compass. Readings are displayed on tuning dials or, for the probe, on a conventional meter.

Inventor Orton has experimented with a prototype of his device and, according to the patent application, has identified 184 elements. Furthermore, he has found that analysis does not require a sample of the unknown—a picture will do. For example, if a photograph of a piece of copper is inserted into the instrument, it will register 29 Ortonaray units, 29 being the atomic number of copper.

Besides it potential utility in analysis, the device also has potential in nutrition. Orton likens the body to a chemical process controlled by a De Broglie waveform made up of contributions from all of the 184 elements his instrument has detected. If the body does not contain all of those elements, according to the patent application, the endocrine glands misfire.

A balanced diet normally will ensure a body burden of all .184 elements, Orton says. Just in case, however, one can supplement the diet with placebo tablets packed with nutrition by treatment with Orton's apparatus. One simply inserts the probe into a container of tablets and leaves it there until the instrument registers 184 Ortonaray units. Regular popping of the treated tablets ensures "full and complete operation of the endocrine glands...leading to better health."

One could equally well jack up an ordinary foodstuff—bread, for example—to ensure a sound diet. Orton prefers placebo tablets, however, because "by using a neutral material such as this...as a supplement to an otherwise carefully balanced diet, mimimum interference in the remainder of the diet will result."

K. M. Reese
Chemical & Engineering News
October 29, 1979; p. 44

The Copieysts Tale...

I have long refused to believe that olde monks actually sat in their cells producing handwritten copies of learned books—there had to be an easier way. After much research, I have found a fragment of a 14th Century document, a document that contains conclusive proof that olde monks were skilled in the art of xerography!

On reflection, the clues abound—terms such as "copyist" and "illuminator" clearly refer to copying not writing. It is my thesis that the original discovery of xerography occurred on some dank, mist-shrouded isle in the following manner . . .

Having spent the morning preparing a sheet of plain white paper, a monk leaves the zinc oxide-coated paper to dry in his dingy cell. After midmorning flagellation, the monk returns to his cell and, following standard procedures, vigorously burnishes the paper, using a skin of a black cat. For fresh ink, the monk collects some lamp-black soot (now known as carbon black) and, to his horror, the monk spills the soot on the freshly burnished paper. "Oh Leviticus!", says the monk, as he vainly tries to remove the soot, but imagine his amazement when he discovers that the residual soot forms an image of his quill pen (which had lain on the paper, as a freak ray of sunshine had crossed the cell).

Doubtless much trial and error lay ahead; but eventually the essential details of xerography must have become clear. Line development must have been well known, but solid areas remained troublesome (witness the hand-coloring in the *Book of Kells,* etc.)

It is now clear why the early religious communities were so seclusive; imagine the secular applications for copying— Royal proclamations, tax forms, rules and regulations, etc. This reticence, of course, led to the loss of the art of xerography—it's not clear whether shortages of black cats, a trend toward the whiter titanium dioxide, or a general improvement in the lighting in monks' cells was to blame for the loss, but clearly the Dark Ages were well named.

For those who might doubt the above thesis, perhaps my commentary on the 14th Century manuscript will allay doubts.

50 And now I wol speke of oure werk,
   As scriven by oure lerned clerk.
   Whan we been there as we shul exercise,
   Oure elvysshe craft, we semen wonder wise,
   What sholde I tellen ech proportion,
   Of thynges whiche that we werke upon.
   As on fyve or sixe ounces, may wel be,
   Of alum or som oother selenie,
   And bisye me to telle yow the names,
   Of orpyment, brent bones, iren squames,

60 That into poudre grounden been ful smal,
   And in an erthen pot how put is al,
   And powdre yput in, and also papir,
   Biforn thise poudres that I speke of heer,
   And wel ycovered with a lampe of glas,
   And of muche oother thyng which that ther was.
   Yet lost is all oure labour and oure hoppe,
   For macheen do maken mooche lite a copie!
   Noght helpeth us, oure labour is in veynen,
   So looketh us for what to blamen,

70 Mowe in oure werkng no thyng us availle,
   For lost is al oure labour and travaille,
   And al the cost, a twenty devel waye,
   Is lost also, which we upon it laye.
   And for all that we faille of oure desir,
   For evere we lakken our conclusioun,
   To muchel folk we doon illusioun,
   And gold, be it a pound or two, borwe
   Or ten, or twelve, or manye sommes mo,
   And make hem wenen, at the leeste weyw,

80 That of a pound we koulde make tweye.
   Yet is it fals, but ay we han good hope,
   It is to doon, and after it we grope,
   But that science is so fer us biforn,
   We mowen nat, although we hadden it sworn,
   It overtake, it slit awey so faste,
   It wole us maken beggers atte laste.

G. Chaucer

Line 57:	While some may claim that this line refers to moon-worship, I feel that it clearly indicates a strong understanding of photoconduction.
Lines 59–62:	This is clearly a recipe for an alleged toner/carrier—as is usual in patent claims, only some of the ingredients are actually needed.
Line 67:	Control of exposure/development was a problem even in the 14th Century.
Lines 71–75:	Research/development always costs more and takes longer than planned.
Lines 76–80:	A good return on investment is always an enticing argument
Lines 81–86:	The step from reduction to practice, to commercial product is never simple.

Robert J. Nash
Chemtech
October 1979, p. 648

Cautionary Tales

Answer

Dwar Ev ceremoniously soldered the final connection with gold. The eyes of a dozen television cameras watched him and the subether bore through the universe a dozen pictures of what he was doing.

He straightened and nodded to Dwar Reyn, then moved to a position beside the switch that would complete the contact when he threw it. The switch that would connect, all at once, all of the monster computing machines of all the populated planets in the universe—ninety-six billion planets—into the super-circuit that would connect them all into the one supercalculator, one cybernetics machine that would combine all the knowledge of all the galaxies.

Dwar Reyn spoke briefly to the watching and listening trillions. Then, after a moment's silence, he said, "Now, Dwar Ev."

Dwar Ev threw the switch. There was a mighty hum, the surge of power from ninety-six billion planets. Lights flashed and quieted along the miles-long panel.

Dwar Ev stepped back and drew a deep breath. "The honor of asking the first question is yours, Dwar Reyn."

"Thank you," said Dwar Reyn. "It shall be a question that no single cybernetics machine has been able to answer."

He turned to face the machine. "Is there a God?"

The mighty voice answered without hesitation, without the clicking of a single relay.

"Yes, now there is a God."

Sudden fear flashed on the face of Dwar Ev. He leaped to grab the switch.

A bolt of lightning from the cloudless sky struck him down and fused the switch shut.

Frederic Brown
Angels and Space Ships
New York, Dutton, 1954

Views In Verse

The Poison Ivy
(A Song of Elizabethan New England)

The woods are fragrant in the spring
With scent of leaf and blossoming:
The trees are all with catkins hung,
And purple orchids bloom among
The poison ivy.

In summer, when the days are long,
The woods resound with thrushes' song.
Sweet honeysuckle from the hedge
Goes trailing to the river's edge
With poison ivy.

The grapes are ripened in the fall
Along the bushes by the wall.
The maples blush before they shed,
And all the hills are flaming red
With poison ivy.

The woods succumb to winter's thrall,
And snowy blankets cover all.
We tread secure in heavy boots,
But 'neath the snow there lurk the roots
Of poison ivy.

Keep It Clean

Don't put germs on the moon, boys—
Don't go and sully her face.
There's too many firms
Of terrestrial germs,
So let's not contaminate space.

There's microbes all over the land, boys—
There's bugs in the oceans as well;
And if somebody soon
Goes and mucks up the moon,
We'll have nowhere that's sterile but Hell.

Group Selection

The bee that stings a man thereby
Condemns herself to pine and die.
What benefit can be implied
By such an act of apicide?
It helps, in fact, to keep alive
Her little sisters of the hive
Whom we might swat—as is our wont—
But, having learned our lesson, don't.

You Wait!

You treat us bugs
Wilth rays and drugs
In quest of morbid sports,
To grow awhile
In dungeons vial
Or flasks of diverse sorts.

You probe with blocks
In mutant stocks
To solve your victual questions;
Then feed us breis
To neutralize
Our inborn indigestions.

On milk, or meat,
Or some 'complete,'
Our cultures thrive—in vain;
For o'er the brink
Of yawning sink
You flush us down the drain.

The time, of course,
Will come, perforce,
When *you* will meet *our* terms;
But until then,
Goodwill to Men.
Sincerely yours,

—The Germs

Relative Speeds

The bee, that zooms from hive to flower
At seven millions lengths per hour
Can spare no time for mice, that go
Five hundred thousand lengths or so.

But mice run ten times faster than
The best laid schemes of modern man:
Some fifty thousand lengths we get
In racing car or flying jet.

While Bannister, when sorely pressed,
Does 15,000 lengths at best.

The tortoise, when he hits the trail,
May touch a thousand, like the snail;
But this is lightning—for it means
Eight times as fast as ocean Queens.

Our Earth, with all its earthly power,
Goes hurtling through the heavens vast
At seven lengths within the hours.
(A bee's a million times as fast.)

Ralph A. Lewin
The Biology of Algae and Other Verses
University Press of America
Washington, DC, 1981

Joseph William Mellor, FRS (1873–1938) made various innovations in ceramics and wrote voluminously: a 16-volume treatise on inorganic and theoretical chemistry, textbooks on chemical statistics and dynamics, higher mathematics for students of chemistry and physics, and quantitative analysis with emphasis on silicates and related minerals.

He also wrote and illustrated "accounts of trivial or passing events ...for the amusement of myself, my nephews, my nieces, and my friends— young and old." Some were letters written to children just learning to read, so the pieces quoted here are not entirely typical of the collection.

Bad Observations—The Enemy of Truth

They darken truth by fancies.—**M. F. Tupper**

I know a man whose main work is in science, but whose hobby is spiritualism.

He has written the story of his conversion to this cult. A recognized medium stayed at his house for a fortnight, and then gave a demonstration by going into a trance; and while in that state at a seance, she told him things about something or other that he said no one but he and his wife knew. This seems to have made a great impression on him, because he concluded that the medium must have acquired the information "supernormally." This word, by the way, may or may not be needed; I have heard it called an evasion of the well-known English equivalent. Of course, such differences of opinion turn on the meanings attached to the words *natural* and *normal*. Before making a supernormal inference of such momentous importance, a hard-headed business person would want to know the thousand-and-one subjects of conversation between the mediumand the wife during the preceding fortnight—not so with this willing believer.

The medium afterwards gave an illustration of planchette writing, and the neophyte stated that at one stage of the exhibition, the guiding hand paused in the attitude of listening. Surely, this is mixing observations and opinions rather badly, a thing the adept would not tolerate in his or her own special department of science. I have tried and tried again, without success, to place my hand in the attitude of listening.

Still further, the medium gave a demonstration of table-turning, and the proselyte wrote that the table appeared to be very pleased with itself, and shook as if laughing! I have walked round and round the dining room table trying to picture how it would look if pleased with itself, but the attempt was a complete failure; and as for laughing—not a smile!

Surely, these are glaring examples of one of the deadly sins in making accurate observations—mixing opinions and facts. The old proverb: "Seeing is believing" is inverted, and believing is seeing. What we wish, said Demosthenes, that we believe; and what we expect, said Aristotle, that we find. The protagonist is continually warning us against the perils of malobservations, and yet, as you will gather from what has just been stated, he can be a very grievous sinner in his dealings with spirits. This incongruity has been noted by others. Can it be that he has a clearer vision than many of us because

His talents of a different nature are—
Not for the daily intercourse of life,
But for the higher things?

J. Baillie

It is safe to say that I should be strongly inclined to accept any statement he made about his observations in his own special department of science; but after what I have seen of his writings, I should be equally strongly inclined to hold in suspense any statement made about his observations in the spirit world. There are many other examples of similar "singularities," thus, Isaac Newton wrote his *Principia* which, considering its day, was one of the "brainiest" books that the world has seen; but he also wrote *The Prophecies of Daniel and the Apocalypse of St. John,* which is in very great contrast with the *Principia.* Can it be that this collection of nonsense is a mild example of another kind of aberration?

Again, we are told by one writer, in a main clause, that "if we really believe...we shall not be incredulous about evidence for supernormal facts...." Can this statement have been made seriously? If we really believe, we shall not be incredulous about any of the statements in this book! Have I been wrong all these years in trying my best to avoid that very attitude of mind when attempting to examine evidence critically? Are not oracular professions of this kind worthy of the fate of the Dedonic Zeus—an infinity of oblivion? Are they not worse than passively useless? Are they not positively mischievous? Is it not better far to obey the Delphic injunction, "Friend, know thyself," or, to pray with the immortal Burns:

> O wad some power the giftie gie us
> To see ourselves as others see us?

It is a curious thing that, to an outsider, many expositors appear almost to have more confidence in their inferences about the spirit world than they have about their experiences in everyday life. The past history of the world, however, shows that it is a general tendency of the human mind to be most cocksure where positive facts are meager. Is it conceivable that Francis Bacon was right when he said: "Truly in all superstition, wise men follow fools"? I can draw two quite different meanings from this quotation.

I cannot understand why some amateur spiritualists are so fond of table-rapping when M. Faraday's instrument—described in his *Researches in Chemistry and Physics* (1859)—shows definitely that the muscles of the sitters always move *before* the table, making it appear that unconscious muscular movements are alone responsible for the bewitching of the table. So far as I know, no one has yet shown that Faraday's demonstration is invalid.

It is almost profane to diminish the charm and romance associated with table-rapping by the application of cold, remorseless facts; but to import supernormal or supernatural agencies to account for the antics of the table when an adequate, positive, normal, and natural explanation is available does not attract the seeker after truth. There are also W. B. Carpenter's observations (1852) on the influence of suggestion in modifying and directing muscular movement independently of volition; and J. Tyndall's experience (1879) at a seance where he gripped the table leg between his knees, and in the subsequent tug-of-war, *pull spirit, hold muscle,* the leg muscles won every time. I have never heard whether any one has been sufficiently interested to try Faraday's instrument with planchette writing; or to try to see whether an alleged spirit could evade an expert lasso-thrower.

I am not here discussing spiritualism *per se;*

> I come not here to be your foe,
> To curse and to deny your truth.
>
> —**M. Arnold**

My purpose is to emphasize the special need for care in mixing the facts with our opinions about the facts. I have done this be exhibiting to you flagrantly bad examples, just as the Spartans demonstrated the evils of intoxication by holding parades of drunken men.

The Whisky-and-Soda Fallacy

In connection with your little attempt to emulate Monsieur Lecoq, the detective, in dealing with the case of stealing you mentioned in your letter, you have based your inference on what is known as the first canon of inductive logic: "The sole, invariable antecedent of a phenomenon is probably its cause," but that principle can easily lead you astray, as it has done many a writer of detective stories.

Some time back, I had occasion to show that the resulting fallacy was prevalent in industrial work in attempting to locate sources of loss. I was quite unable to make my meaning clear until I called it the "whisky-and-soda fallacy," and illustrated the point by a very old "wheeze." A lush gets charged up with whisky and soda water on Monday and becomes drunk. The lush wonders whether it is the whisky or the soda water that brought on the inebriation. Our tippler therefore tries brandy and soda water on Tuesday and is drunk again; while on Wednesday the sot tries gin and soda water and is drunk again. The facts are here summarized:

Beverage	Effect ("soda" constant)
Whisky and soda water	Drunk
Brandy and soda water	Drunk
Gin and soda water	Drunk

Since the sole, invariable antecedent of a phenomenon is probably its cause, it was inferred that the soda water was the cause of the drunkenness. Obviously, the soda water is not the *sole* antecedent, because *two* constant factors are involved—the soda water, which is obtrusive, and the alcohol which is concealed. Of course, in this example, if soda water alone had been tried, it would have been eliminated, but, in your case, the rule is a useful help because it was not physically possible to isolate one of the factors directly. Your conclusion may be perfectly correct, and yet, to make it convincing, you must eliminate the possibility of the presence of other constant factors vitally important, but not so obvious. All the same, I rather like you paying first attention to the obvious factor. Now give your conclusion a very severe twisting; in fact, as a sage has said, it is a good plan always to look very carefully a second time into conclusions that you were cocksure about the first time. There may be a granny's knot in your argument. One who knows but partly can judge only in part.

The Utility of Scientific
Research—Iconoclasm

Most people now admit that scientific research is a necessity, and not a luxury—when taken industrially, and in every other way. Many a time, when I have not been able to pass definite data on to the thinking-box, the brain has positively refused to do its job. It is one of the functions of research to find these data. Few people can fail to realize that though the relentless march of modern research has placed in human hands new powers that "transcend in strangeness and grandeur the wildest fables and dreams of antiquity," it has also altered, and sometimes reversed maxims and usages that are the inheritance of centuries. Even with some of the most familiar subjects, the data seem to have got topsy-turvy, for, in many cases, as I have just indicated, we have been blinded by their familiarity. For example, from the time we were in short trousers, we have grown up with the idea that "A watched pot never boils." An investigation, reported by a writer in *Punch,* tested this axiom in several ways, so that in one set of observations the pot knew that it was being watched, and in the other set, the pot did not know that it was under observation. No matter whether the pot was watched by someone with normal eyes, or with cross-eyes; no matter whether the watching was done by someone standing in front of the pot, or by someone hidden from the pot but peeping at it from behind a screen, the pot always boiled. Ergo, *a watched pot does boil.* Thus scientific research has shown very definitely that the ancient adage is wrong.

At first blush, this result might evoke an ironical reference to the venerable admonition:

> Teach not a parent's parent to extract
> The embryonic juices of an egg by suction,
> The good old lady can the feat enact
> Quite irrespective of your kind instruction

or to Terence's *Heauton:*

> For he, in truth, with much ado
> Would prove that one and one make two.

Of course, some maxims are truisms; they are bald statements of facts, and "facts are chiels that winna ding." For example, the old proverb, "looks cannot kill," must be a truism, otherwise it would be suicidal for some people to look in a mirror.

J. W. Mellor
Uncle Joe's Nonsense
for Young and Old Children
London, Longmans, Green Co., 1934

On Making Rounds

In view of our chronic concern over the adequacy of teaching programs for the production of well-rounded pediatricians, it behooves some of us to expend a reasonable effort on explaining the art of making rounds.

Directors of medical education are becoming more and more aware of the gaps in the teaching of many of the practical and important facts of medical life. For example, what instruction is given on a foolproof way to keep oral and rectal thermometers separate? Will any new graduate from a residency completely understand the subtle art of income tax deductions? How can one develop the knack of conversing with a neurotic mother over the phone while simultaneously finding storage room for the latest unwanted batch of samples? How does the odor of a stool on a diaper lead to a major change in diagnosis? Obviously, as these and other scraps of potentially useful information must and will be taught, it becomes inevitable that even less time will be left available for actual bedside teaching of the art of making rounds. It is hoped, therefore, that this guide will be helpful to residents and others in identifying the various categories of experts in the field of roundsmanship.

Syndrome Man

Perhaps the easiest character to recognize is the "syndrome man." Because he usually is not bashful he can be spot-

ted promptly.   Once, early in his career, he overheard the
chance remarks that "you cannot treat it until you diagnose it."
Deeply impressed, he began and still continues to accumulate
endless volumes of notes and diagnoses on the assumption that
diagnosis will solve all problems. He is first heard from early
in the year when the assembled staff finds a child who has ears
which stick out, a pointed head, and a beak-like nose. Further
examination reveals sticky rales in the left base posteriorly,
strabismus, an undescended right testicle and a clear corrosive
discharge from the umbilicus. Obviously this is a syndrome.
Everyone looks at the Chief, who says, "Wish Steve were
here; he could name this immediately." Had Steve been there,
the eponym would have been dropped long before the Chief
could clear his throat. If Steve does not come to light in this
fashion, he may still be identified by his casual approach to the
art. Should a perfectly straightforward case of fibrocystic disease be under discussion, the "syndrome man" remarks how
much this child's ears remind him of little Hortense Slup, "who
had Hossenpfleiger's disease several years ago. This move is
expert in every sense. There is no counterploy, but here it is
interesting to watch the other staff members. One can here
learn the discreet smirk and most of its variations. A note of
warning: Don't challenge the "syndrome man." He knows
what he is talking about whether it means anything or not. In
all fairness, however, he proves a most valuable staff member
when visitors from Boston or Baltimore arrive. The Chief

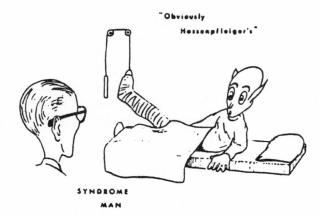

"Obviously
Hossenpfleiger's"

SYNDROME
MAN

knows that he can suggest a quick (two hour) tour of the wards with the complete assurance that none of the visitors will be able to outsyndrome his group. No conference champion ever lacks bench strength.

Paper Quoter

Considerably more common, almost as easy to spot, and perhaps a bit more useful is the "paperquoter." He is not unlike the sparrow to the bird watcher; he seems hardly worth recording after the first few days. However, his inherent danger makes quick identification mandatory. There are several varieties that can be conveniently grouped according to time. The young man usually starts out as a "current paper-quoter." He comes to rounds armed with the absolutely latest published paper on several possible subjects. He has memorized the last name of the senior author (the other authors become "and his group"), the color of the paper binding the journal, the year and month of publication, and the conclusions. All of this he rattles off at the slightest stimulus.

This man is vulnerable and there are several kinds of counterploys. Any one of these may be tried without significant risk. The easiest and possibly the best approach is to suggest immediately that the data do not show significant differences. (One may add here, "at the 5% level" if he feels gay at the moment.) Since the "paper-quoter" has no idea what the data are, he will reply that, in his opinion, "they looked convincing." There has been a step down from the author's opinion to a personal opinion and a goal has been scored.

A similar score may be achieved when a paper on treatment is quoted. One may say simply that the study was not controlled (this leaves one open to the accusation of withholding life-saving treatment from desperately sick children). Or better, one can say the test was "not well conrolled or randomized, using the double blind technique." This is always good.

Finally, one can challenge with the statement that the experimental groups were not comparable. This is especially recommended for the young roundsperson. It has several points in its favor. The "paper-quoter" will have to report how many groups were submitted to scrutiny (the more groups, the greater the likelihood that they are not comparable). Fortunately, the challenger can get along with absolutely no knowledge of the paper itself. Since the paper is current, there is at least a 99 percent chance that it is worthless and at least a 90 percent chance that the specific criticism will be valid for that particular paper. That no preparation is required is handy for the young house officer who has all that can be handled with the chief resident, the records librarian, and personal sex problems (almost always three separate and distinct entities).

Another word of caution: don't be greedy. The "paper-quoter" may be challenged safely in such a way not oftener than once every four to six weeks. Short-term situations may find the quoter well loaded on all the current favorites. To accurately judge the waxing and waning of real knowledge involves some trial and error, with an occasional skinned nose, but the results can be extremely rewarding if one takes the long view. A final warning: be sure you have identified the right person. The gambits mentioned are based entirely on the supposition (usually correct) that extremely superficial knowledge is enjoyed by the quoter. Fortunately the opposite is rare.

Frequently (perhaps one should say: not infrequently), the "current paper-quoter" goes on to become an "old paper-quoter." This occurs because the quoter is vulnerable, responding to this state by discussing only those topics on which our citer would like to be considered an authority. Fortunately, here inherent weakness comes to the surface; the quoter still selects far too many topics. For each topic our quoter mentions an article about five to eight years old of which once some favorable remarks were made by respectable company. Thus

armed, the quoter enters the fray, but of course, is easily identified by the dates of the papers. Once identified, the counter-
attack becomes a bit more difficult, but much more interesting.

One of the most intriguing moves is the Penny Ploy. This
was originated by Marcus Penny for use when the "old paper-
quoter" comes up with a very ancient selection. The victim
goes into the usual routine, giving author, journal, and date.
Our hero then asks, with a perfectly bland expression "What
page?" Since no roundsperson worthy of the name can ever admit ignorance, the quoter replies lamely, "Page 324, I think."
Obviously, our citer does not know, and better still does not
know what has happened to him or her, until discovering blood
on the back of his or her shirt. This is great fun!

A more difficult approach, although perhaps less fun, is
the "I've read that article also" gambit. This obviously requires
preparation and patience, but the "old paper-quoter" can be
expected to go through all of the repertory at least once and
then start all over again. One begins by quoting any part of the
paper not mentioned in the opening statement, and then chal-
lenges the conclusions. Whether or not the challenge is well
founded is unimportant since doubt has now been cast. This
can be very effective, especially against a relatively recent
arrival from the East.

Sometimes a "current paper-quoter" may advance to a "personal communication citer," now stating "Emmett would not
agree with that," or "Jimmy told me last week that they were
using distilled water and getting very good results." This can
be an effective move, hard to fight. To keep this citer at home
and not traveling is difficult, but may be the only real hope.

Name Dropper

Even more difficult is the roundsgoer who becomes a
"name dropper." This type is a variation of the "personal
communication citer" who is attracted to the cult of personality
and with very little encouragement could have become a
Ghandi follower or a Henry Wallace devotee. The namedropper has been around, having spent at least a month in some capacity in the shadows of at least five or six well-known people,

all of whom are now presented as first name friends. With the name dropper, discussions will be spiced with "Well, Ralph always said that aspirin is a good drug," or "Well, Jim always told us that meningitis is bad." The reader will immediately sense that because Ralph is said to have said something 80 times does not necessarily make it so, but the glamor of the name is difficult to outweigh. At least two defenses are open. One is to wait until the Great Leader is quoted out of the usual field of expertise. Since most quotations are more or less innocuous; this requires patience. When the time is ripe, a direct challenge is sufficient. This always increases the anger of the namedropper since Ralph can never be wrong. However, it may effectively silence our idolator for several weeks until emotions are under control.

"NAME DROPPER"

The other defense is to fabricate thus: "Yes, I heard our Nobelist say that in 1932, but that opinion has been changed at least three times since then." Again doubt has been cast and emotions aroused. This is sufficient.

One almost impregnable type of "namedropper" is the "night-spender name-dropper." Women are masters at this. On every available occasion they gravitate toward each other

like two Americans in Moscow. This usually ends with their
staying together for a night, talking cozily. At that time, after
having discussed corset, abdominal scars, and former fellow-
interns, they brief each other on their own "current work."
Little of this work is ever published, but it does get wide
circulation in this way. There is little that one can do here
except be satisfied with a masculine smirk. Also, you may
hope that one of the principals will get pregnant, which will
take her out of circulation for a time. Unfortunately, one sus-
pects that the latest notions of contraception have also been
exchanged at that midnight tete-a-tete.

A most difficult problem is the husband and wife team.
Each large center and almost all small pediatric groups have at
least one such team. To insure a marital truce, if not connubial
bliss, they have adopted somewhat separate specialties within a
broad field. Thus complemented, they do not hesitate to enter
the fray. Margaret speaks until the Chief begins to frown. She
realizes at once that she is on thin ice and passes the ball to
Percival with the statement that she saw him reading an article
last night in the CMJ *(Calcutta Medical Journal)* about which he
should speak. He does. She smiles and regroups. The Chief
yawns; at which point Margaret starts off again. Finally one of
the pair states: "We both think...." This is the signal for the
Chief to walk out of the room with the retinue stumbling behind. The "we both think..." atmosphere effectively keeps
down the number of applicants for full time staff positions.
Thus, the team is very popular with the Dean. One soon
decides that there is little to be done about it and resolves to co-
exist.

Bed Identifier

An interesting specimen is the "Bed-Identifier." Someone
mentions Imogene Gluty, who had Letterer-Siwe's disease and
was in the hospital several years previously. Immediately,
someone else claims to remember her and that she was in "that
bed right over there." This seemingly simple statement impresses the junior members of the staff immeasurably and the "Bed-
Identifier" basks in the glory of an apparently limitless memory
for information.

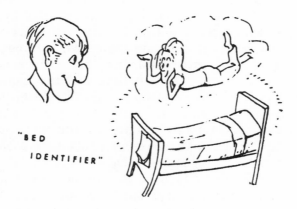

"BED IDENTIFIER"

Equally interesting is the person with the cryptic question. This doctor is definitely looking for trouble and may or may not have any pertinent information to contribute, usually not. The cryptic-questioner can be and frequently is the counterpoise to the "namedropper." The "namedropper" may say: "Well, Ignatz Minsky always said that..." at which point our friend says "Why?" Since the "name dropper" had never stopped to consider why Ignatz ever said what "he always said," the confusion resulting from this collision can be quite gratifying.

Walking Summary

Another individual of note is the "Walking Summary of What We Have Had on Our Own Service." This title is self-explanatory. The Walking Summary is completely indispensable to the Chief who will want to know: "Les, how many cases of this do we have in our files?" Les, without hesitation, replies: "We have 82 cases on record of which 31 had splenectomy, 8 died, 52 are well, and 12 still have symptoms." Les usually makes all of this up. Even so, the Chief understandably beams and this is how the academic world has become studded with Associate Professors with this sort of a background. It is good to be friendly with such a person, who may well be Dean some day.

"Walking Summary"

The Cusser

Another variety is the "Cusser." Shockingly enough, this is occasionally a female who responds to any challenge with: "Hell, I *know* that's so, Goddamit." Delicate ears and not a few sensitivities are thus offended. Most roundgoers tend to lapse into silence. There is no treatment except infinite patience.

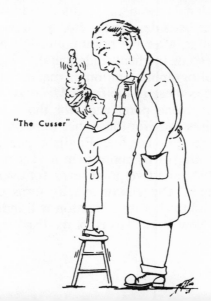

"The Cusser"

Old Timer

Quaint is probably the best word for the "old timer," whose basic qualifications include gray hair and years. Some degree of deafness is of value. With the old timer the introductory remarks are basically deprecatory, starting out with "Well, you folks know a lot more about this than I do but . . . " which is then followed with ten minutes of solid information spiced with clinical perspective and well-conceived critique of the one-gene, one-enzyme hypothesis. This individual is to be fed and watered with care.

The "place-quoter" is a variety of "name-quoter" and has many of the same qualifications. However, a word of caution here, because all geography is relative. To say "When I was in Boise...." will impress few in Miami, whereas the word "Boston" can call forth gastrointestinal or vasomotor responses in almost any setting. Unless one is well established, this approach carries the risk of evoking an invitation to return to the source, and should be used only with forethought and judgment.

Sometimes a visitor comes to rounds who brushes aside the pecking order and within five minutes tries to show up all the namedroppers and the paper-quoters. References and quotations start being batted back and forth with dizzying rapidity, though always with politeness and sometimes with humor. Loose feathers swirl more and more wildly until the Chief, noting that the locals are tiring, calls it a draw and leads the group to another bedside. The victor of such an exchange is decided upon later, after the point count has been computed. There is always some medical student at the periphery who jots down every bibliographic citation as made. Later in the day, our student checks these in the *Index Medicus*, and reckons the score as follows: one point each for those items that were correct; two points each for those that were off by more than a year or in a language other than English; and five points each for those that were off by more than a year or in a language other than English; and five points each for every one that was wholly fabricated. The visitor usually turns out to have the highest score, and further investigation will indicate that training occurred in New York, Texas, or the Far West.

The "question–answerer" tends to think in tight ever-decreasing circles. Since this sort is thought to be an authority is something or other, an occasional question is directed to the answerer. Usually, the question has nothing to do with the answerer's field, or is one for which our expert has no answer, or both. The answerer's response is to present data on another unrelated point, with firm conclusions. To profit from the discussion one must quickly think up the question for which the answer is now being given. It is usually at this stop in the rounds that the "old-timer" inspects both hearing aids carefully, convinced that something is being scrambled somewhere.

OLD TIMER

Many other varieties and subvarieties of roundspeople are to be found in nearly every habitat. Needing more intensive scrutiny are such variants as the electrolyte-oriented, the idea-stealer, the psychobiologist, the dose-questioner, the no-sayer, the convention-goer, the nonlistener, the otoscope-dropper, the early-arriver, the anti-antibiotician and the pro-prophylaxer.

It is hoped that this analysis will encourage more research in this area. To dissociate genotypic traits from adaptive coloration and from pretension wearing the garb of expertise may take some time, but is worth the effort, though it may leave one wiser and sadder.

It is emphasized that these data are "preliminary," and that "much more work would be desirable." There may be very fundamental principles in this field whose existence has thus far not even been imagined. On this basis a grant application is in process of being drawn up.

J. Earl Roberts
Clinical Pediatrics, 3
No. 6, June, 1964, pp. 370–373

The mailing wrapper of a computer catalog carried this warning notice:

If you throw this in your wastebasket unopened, a capsule of water inside will break, spilling onto a dehydrated gorilla. He will then jump out of the envelope and hug you to death.

A similar mailing carried this sardonic commendation:

> **THANK YOU FOR SMOKING!**
>
> YOUR EARLY DEMISE
>
> WILL HELP STABILIZE SOCIAL SECURITY

Parody

Geolimericks

A geophysicist named Ray
Said "I must make my PhD pay,
An orogenic crust
May be a bust,
But at least now I publish, OK?"

As granitic crusts are afloat
And basalt can act like a boat
We're all up a creek
If they should spring a leak.
New line geologists, take note!

Tectonic plates are news, though
A. Wegner just always knew so.
The transform fault
Made critics halt,
And the theory grew, Tuzo!

The Moho was quite justly peeved,
Of a function it has just been relieved,
For geologists have found,
That the plates grind around,
On an asthenospherical rheid!

A. L. Taylor

269

In matters preceding orogeny
It is difficult to follow ontogeny.
It seems like subduction
Is tectonic seduction,
And mountains are merely the progeny?

David Rostoker

There once was a land named Gondwana
That had Glossopteris and fauna.
It broke with a rift
And started to drift,
Today it's known as Botswana.

International Stop Continental
Drift Society Newsletter, vol. 3, no. 3
January 1981, pp. 6–7

The Auklet is the parody version of the American Ornithologists'
Union respected quarterly, The Auk. It has appeared intermittently (41
issues from 1920 to 1982), often at the annual dinner of the AOU. Early
issues contained references to members's personal foibles and papers
mimicking the writings of specific writers. For example, the research of
Charles Sibley was featured on two covers (1965, 1974), his style is
captured in "Electrofreezing..."

Electrofreezing of Egg Yolks—The New Final Answer in Taxonomy

by Chick Giblet

Birds have always interested humankind, and from the
earliest days humans have attempted to set up classification of
the known avifauna *(Genesis, II: 19–20)* {Ed. note—five more
pages of this sort of padding here removed to get to the point}.

Recent experiments on our big expensive machines have indicated that all previous final answers to bird taxonomy have been outmoded, even the ones we came up with using our *other* big expensive machines. Yolks of the eggs of 1066 species of birds, representing 15 orders (as of the new outmoded classification; three orders from now on) (*see* Appendix I for list of species) have been secured from a variety of sources (*see* Appendix II for lists of sources), and have been tested by our corps of special technicians in our new extra special machine (*see* Appendix III for list of foundation grants). It has been found that when avian egg yolks are subjected to temperatures approaching Absolute Zero, the pattern of ice crystals formed is absolutely diagnostic for each taxon when examined through our special expensive ultralow-temperature polyphase three-cushion microscope. If the crystals look alike, the birds are closely related; if they look different, they are not closely related. At present we have no strictly objective definition of "alike" and "different," but you can take my word for it. A summary of some of our preliminary results is given below (*see* also Plates I–XXIV):

ORDER I (name to be determined by specialists in priority, not me): *Characterized by:* simple, unbranched crystals clumped into aggregations of various sizes and shapes (*see* Plates I–VIII). *Included groups:* Loons, grebes, penguins, diving ducks, cormorants, four-toed woodpeckers, Carolina chickadee, tinamous, hjornbills, and Nehrkorn's blue-tailed slyph *(Neolesbia nehrkorni).*

ORDER II: *Characterized by:* crystals with simple branching, singly or in clusters (see Plates IX-XVI). *Included groups:* Albatrosses, pelicans, dabbling ducks, anhingas, three-toed woodpeckers, wrens, toucans, rheas, guineafowl, oxpeckers, adult honey-guides, and Golden Seabright Bantam.

ORDER III: *Characterized by*: crystals with compound branching (*see* Plates XVII–XXIV). *Included groups:* Ostriches, thrashers, owls, black-capped chickadee, weaver finches, cranes, juvenile honey-guides (artificially induced ovulation), hawks, oropendolas, soft-shelled turtles, and the Comte deParis' Star-frontlet *(Coeligena lutetiae).*

INCERTAE SEDIS: White Leghorn. Additonal species are now being tested, and new versions of this classification may be expected from time to time.

NOTE: This particular Final Answer has a unique inherent advantage in that the test materials are not damaged or spoiled. When somebody comes up with a newer final answer, these test materials can be salvaged for an excellent omelette (*see* Appendix IV for recipes).

> **Kenneth C. Parkes, Editor**
> **Richard C. Banks and Robert W. Storer**
> in *The Antic Alcid*
> *An Anthology of the Auklet*
> The American Ornithologists' Union
> New York, 1983

Old Twists to an Old Tale

Under this title, Jerry Donohue reviewed The Double Helix: Text, Commentary, Reviews, Original Papers, *edited by Gunther S. Stent (Weidenfeld & Nicolson/W. W. Norton, 1981). In his comments on the controversial history of DNA, he delt with The Myth of the Invention of Base Pairing, The Myth of the Undercover Agent, the Myth of the Filched Photo, etc. Boxed in Donohue's paper is a review cast in the style of the Kalevala, the Finnish epic poem the style of which was later adopted by Longfellow).*

Review of James Watson's *The Double Helix*
By J. Field

Hear the song of how the spiral
Complex, twisted, double, chiral
Was discovered. Its construction
Following some shrewd deduction
Was accomplished. From all sides "Oh
Tell it as it was" they cried so
Then he took his pen and paper
And composed the helix caper
Writing of the Cambridge popsy
Turned the world of science topsy
Turvey like had not before been
Done. The story is no more than
How the structure was discovered
Secret of the gene uncovered

Biological prediction
Better than a work of fiction
Stylistically breezy
Everything appeared made easy
And it was. The bases' pairing
Found at last—a small red herring
Notwithstanding. How he found them
And the hydrogens that bound them
Was a stroke more accidental
Than a work experimental
Chemistry was not his calling
He had read *one* book by Pauling
Models of the bases he'd made
With them on his desk top he'd played
Shuffling them and putting like with
Like he made no lucky strike with
Guanine, thymine, adenine. A
Letter came from Pasadena
Horrors! then the day was won for
Pauling's model was quite done for
Extra atoms he had set in
Where no atoms should be let in
With some phrases less than modest
Pauling's model's called the oddest
Back now to the basic pairing
How's the structure building faring?
Faster, faster goes the race, then
Everything falls into place when
Tautomers which nature chooses
Are the ones our author uses
Two chains (but here we can't be sure)
Plus this extra structure feature:
Pauling's outside bases inside
Pauling's inside phosphates outside
Now they all were quite ecstatic
Soon became they quite dogmatic
Having nature's secret later
Checked against the X-ray data
This was difficult, for Rosie
Found our scientist rather nosey

Bragg compares the book with Pepys. Is
This the verdict of Maurice? His
Thoughts alas we cannot gainsay
Trumpet blowing's not his forte
Tales like this do have a moral
Whether printed whether oral
Find the right man to advise you
Then you'll get a Nobel prize too.

Jerry Donohue
Nature
Vol. 290, April 1981, p. 649

Rasputin, Science, and the
Transmogrification of Destiny

A paper of this title purporting to be a Princeton University preprint was circulated in the early 1970s among relativists. The paper is an immediately recognizable parody of the inimitable style of John Archibald Wheeler (then at Princeton). The paper was eventually published in the journal General Relativity and Gravitation, *5, 175–182 (1974). The paper also received some notoriety when Freeman Dyson reviewed (in* Science *185, 1160, 1974) an actual Wheeler paper with the comment:*

"Recently he (Wheeler) published an article entitled *'From Mendeleev's Atom to the Collapsing Star,'* which inspired some anonymous joker to circulate in Samizdat a parody called *'Rasputin and the Transmogrification of Science.'* The parody was so good that many of us suspect it could only have been written by Wheeler himself. (It was written by William Press, Harvard College Observatory.) Wheeler's chapter in this book might also have been written as a self-parody."

"Wheeler's" paper is too long to reproduce here in its entirety and it resists condensation. But this is the way it starts:

Rasputin's Goal: Not Much Out of a Little

This centenary of the birth of a great scientist–statesman is also by a coincidence the year of the death of a great work, striving for simplicity and unity, for function and wholeness, for design without design, in his famous motto "Less is more." What shall be our motto for Grigory Efimovich Rasputin? He sought to the end to see the fantastic wealth of scientific facts all as transient consequences of one Universal Principle of Vacuity. How much more briefly can we state his theme than "Not Much out of Little." And if the mighty twin worlds of Science and Politics ever float a flag over their long-allied forces assembled in serried rank, what happier motto could they find for it than *Multum Non Ex Parvo?* No words would do greater honor.

Transmogrification of Destiny: A Lot of Nothing

If "Not Much out of a Little" epitomizes the triumphs of the past, then also its immediate antithesis "A Lot of Nothing" or *"Multum Nihilo"* summarizes the present crisis that one can name in the history of the universe: The transmogrification of Destiny.*

Figure 1 on the next page illustrates schematically the dilemma. All flesh is grass, all grass is dust. The details have vanished: *"Ubi sunt.."*

A Black Whole Has No Peculiarities"

The transmogrification takes place on a characteristic time scale. For a destiny comparable to that of humankind, this time is less than a millisecond. Let the universe have a destiny.†

---

*For an account ot the history of the universe, including a summary of the pioneering contributions of Alexander the Great, Gaius Julius Caesar, Attila the Hun, Henry VII, Athelstan Spilhaus, and others, see for example H. G. Wells, *An Outline of History.*

†Man is the measure of all things. (*Ad Hominem metric est.*)

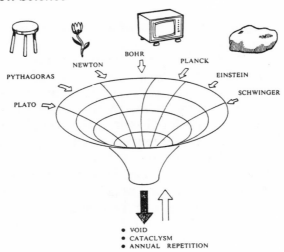

Fig. 1. A black whole reduces whatever matters in particular to a system of void, cataclysm, and annual repetition, thus transcending any particular conversation, including leprechaun and baron. The Cheshire cat in *Alice in Wonderland* also vanished. Only its smile remained behind. What remains here? Void, first of all. This void subsists on the creative desire as firmly as ever. In addition to void, the transmogrified universe possesses cataclysm and annual repetition.

The effective depth of destiny decreases to half its value in a characteristic relaxation time of less than a millisecond. Dropping to half value, then to quarter value, then to eighth, sixteenth, thirty-second, sixty-fourth, and so on, every deterministic feature erased away to the utmost perfection by the end of a single second.

Drop in a philosopher. The philosopher makes a momentary disturbance on the void of destiny, then also shrinks in cataclysmic annual repetition, and fades away. Drop in scholars of all disciplines and factions. They too are transmogrified until nothing more remains, only certitude. Call the resulting entity a "black whole" and summarize the perfection of its final state. Say: "A black whole has no peculiarities."

John Archibald Wyler
Wilmington Press, 1960
Harvard College Observatory

The Modern Doctor Chemical
(Air, "Modern Major-General," from Gilbert and Sullivan)

I am the very pattern of a modern Doctor Chemical;
I send to all the journals my remarks and views polemical.
I've studied mathematics till I think in terms vectorial
And scorn the plodding soul who seeks for molecules pictorial.
The calculus is food for babes; I love a complex var-i-able
And state a simple law in terms the laymen thinks are terr-i-able
I can talk of relativity and space-time for a month or more
And integrate elliptically to terms the $(N + 1)$th or more—
And yet my hand and mind are seized with palsey and paralysis
When I essay that dreadful task—a chemical analysis.

I've mastered all of physics, wave mechanics, and spectroscopy;
There's not a Planck or Einstein whose sage remarks are lost to me.
I know the private life of each electron 'round the nucleus
Their inner quantum numbers and their interactions dubious.
I outdistance Bohr and Sommerfeld in matters theoretical
Anticipated Goudsmit with electron spins heretical—
And yet my heart is filled with woe and primal trepidation
When in the lab I'm faced with an organic preparation.

I surpass Debye and Hückel in their views interionic
Dumbfound LaMer and Sandved with equations not canonical;
I can compute the entropy from levels in the molecule
And fill the journal page-by-page with integrals symbolical.
I can compute the entropy from levels in the molecule
And fill the journal page-by-page with integrals symbolical.
I calculate reaction rate and heat of ac-ti-va-tion
And any other quantity expressed by an e-qua-tion—
Yet still the old-time chemist shakes his sides with vulgar merriment
When I confess how I detest performing an experiment.

Frank T. Gucker, Jr.
Chemical Bulletin
June 1932

Lewis Carroll wrote a piece he called Hiawatha's Photographing, *142 lines in the easy running meter of Longfellow's* Song of Hiawatha. *It appears that we have below a parody of a parody, likely written by a chemist.*

Hiawatha's Photographing

First, a piece of glass he coated
With collodion, and plunged it
In a bath of lunar caustic
Carefully dissolved in water.
There he left it certain minutes.

Secondly, my Hiawatha
Made with cunning hand a mixture
of the acid pyrogallic,
and of glacial acetic,
And of alcohol and water.
This developed all the picture.

Finally, he fixed each picture
with a saturate solution
which was made of hyposulfite
Which again was made of soda.
(Very difficult the name is
For a meter like the present
But periphrasis has done it.)

Anonymous

The Isolation and Characterization of Plentisillin[1]

Plentisillin was first detected thru its unique property of having no antibiotic effect whatsoever. The writer noticed early last year that the mold on his lunch meat (obtained in a local restaurant) was entirely unaffected by a piece of banana peel that had accidentally contaminated his sandwich. The significance of this seemingly unimportant observation was not lost and he immediately disclosed his discovery to the President of the United States. A preliminary $100,000,000 grant for development research was set aside, the funds to be dispensed by the Navy.

With Dr. Si Lane and other colleagues, the writer quickly set up an intensive research program. The first goal was the isolation of plentisillin, as the active principle was later named, and for this purpose Naval representatives obtained 500,000 banana peels from Brazil, shipping them by plane to a secret laboratory on the 25th floor of the Columbia College of Physicians and Surgeons. The peels were separated from the insides by excising one end of each banana and applying pressure to the sides. The waste was fermented to a liquid of considerable value to the experimental staff.

The details of the work involved in the isolation of plentisillin will be presented in another paper.[2] The method that proved most efficient is approximately as follows: (a) Extract the peels 20 seconds with water, after homogenizing and adding Vitamin D; (b) reflux 64.2 hours with an alcoholic solution of deuterium oxide and a yttrium–ytterbium catalyst; (c) discard the solution, dry at 400°C over concentrated sulfuric acid, and grind the residue with dry nitrogen iodide,[3] extract with fuming nitric acid (buffered with perchlorick acid) and pass thru an activated alumina absorption column. At this statge a preparation (about 50 micrograms from 200,000 peels, of approximately 40% purity is obtained by the evaporation of the eluate. A color test, discovered in this laboratory by Prof. Solomon F. O'Namide, as well as nonantibiotic activity were used to follow the course of the concentration.

The color reaction involves addition of the sample to a freshly boiled solution of ammonium nitrite, slow dilution with 99% hydrogen peroxide, and then refluxing with a trace of manganese ions. After several hours the word "plentisillin" appears in greenish mauve Greek-Pittman symbols on the side of the refluxing flask. One of the observers of the reaction suggested that the active principle be named "plentisillin," and after some discussion the name was gnerally accepted. One of the valuable features of the color test is that close derivatives also spell out their names, making the task of structure analysis considerably simpler.

The final purification step proved to be an extremely baffling problem, in the pursuance of which the resources of 20 industrial laboratories and 2 OSRD divisions were drawn upon. The technique finally found successful consisted of preparing a saturated solution of the crude material in boiling water allowing it to cool slowly. Beautiful triclinic hexagons crystallize out in a few days at minus 40°C.

Pure plentisillin (hereinafer referred to as I) is extremely unstable. Exposure to air gives the nitride (azsillin) within a few minutes. Even traces of carbon dioxide decompose it to a mixture of as yet unidentified degradation products. One of the key reactions is that with krypton, which splits I into two fragments: plentimine and cryptosillin. (The latter spelling is unfortunate, but the darn stuff insisted on a "c" in the O'Namide color test.)

Crystalline I has a refractive index, density, and a color. Under ultraviolet light it fluoresces ultraviolet of the same wavelength. X-ray and electron diffraction measurements indicate a unit cell of 4 x 4 x 4 Angstroms, or 20 x 2 x 1, or possibly some other size. Its absorption spectra in neutral and acid solutions are presented in Fig. 1.

Careful chemical analysis, performed on a 10-microgram sample by Dr. Alvin D. Hyde, indicated the following formula:

$C_{42}H_{10}D_2S_2NO$. The finding of deuterium occasioned some surprise, and Dr. Hyde's analytical results were checked and confirmed with a mass spectrograph. The presence of deuterium is apparently due to the presence of deuteriase, an enzyme incompletely isolated from banana peels.

Distillation of 2 micrograms with zinc dust resulted in degradation to a white cystalline material subsequently identified as fluoranthrene. From this reaction a tentative structure may be assigned to I, as shown in Fig. 2. Confirmation of the heterocyclic structure was later obtained by exhausting methylation of plentimine, obtained by fission with krypton *(see* Fig. 3). The final polymer, polyvinyplenti, is a high molecular weight polymer of the reactive trivalent ion, vinylplenti, (II), and has physical properties suggesting practical application as a plastic. The 4-membered heterocyclic structure, known as plenti, is unusually stable owing to resonance and the principle of homology, whereby the sulfur atom may be considered the equivalent of two carbon atoms.[4] The formulas given here were confirmed with olfactory apparatus by Dr. Bark, of these laboratories.[5]

Cryptosillin has the empirical formula: $C_{37}HSKr$[6] and is probably fairly unsaturated. The exact structure has not yet been determined completely, but on the basis of preliminary studies a remarkably simple synthesis has been developed. Krypton hydrosulfide is heated in a sealed tube with excell Norit at 201.3°C for several hours, dissolved in liquid HF, decolorized by treatment with Nuchar, filtered through Super-Cel, and allowed to crystallize. Very pure cryptosillin is obtained in characteristic and undescribed form.

The physiological properties of I are of considerable interest and 10 large extracting plants are being set up thruout the world by the United States. The absence of antibiotic effect has been mentioned previously. This property has been acclaimed by physicians who have conducted tests on a wide variety of diseases. "Thank God for plentisillin," one prominent official of the AMA exclaimed.[7] "We thought business was dying down."

When an average-sized person runs upstairs 10 flights and is quickly injected with only 2 gammas of I, he exhibits tachycardia, hyperpnea, and a high blood lactate concentration. The significance of these reactions is not yet quite understood.

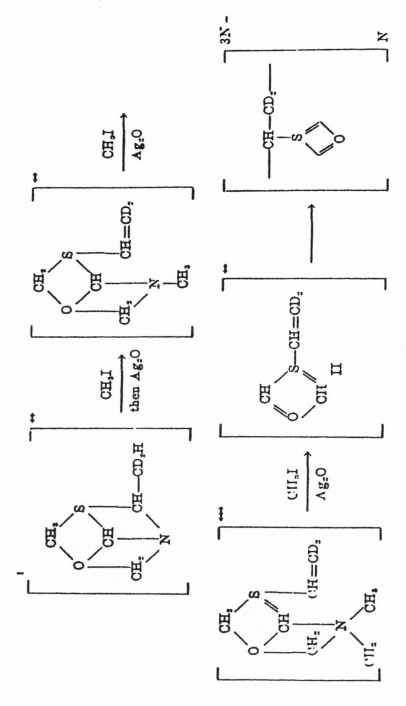

Administered in cases of colonic obstruction and spelled backwards, relief was obtained within several days. One volunteer (from among our staff) apparently developed an anaphylactic response, in which her red cell count jumped to π million, irrational behavior appeared, and she considered changing her profession. Of her subjective sensations she has said only: "Mm-m-m-m."

Some light has been thrown on the metabolic behavior of I by Dr. Hans Bloshemberg, who succeeded in labeling it with radioactive hydrogen (tritium) by catalytic exchange with the deuterium. Ten milligrams were fed to a Hearst special topics reporter who was sacrificed three days later. A detailed analysis of the various organs[8] demonstrated the presence of tritium in the body water and nowhere else. This was interpreted to mean that some of the plentisillin is hydrolyzed by a liver enzyme of oxyhydroxplentisillate, which is absorbed by the spleen and decomposed further to tritium oxide and an intermediate used in hemin formation. The remainder of the ingested I, it was adduced, combines with a globulin to form an antienzyme which catalyzes the nonreaction of tritium oxide with carbon and nitrogen atoms in the *milieu*. (The assumption was made that the reporter had a *milieu*.)

While considerable additional work has been performed on I, this information is still restricted and may not be revealed until published in *Life* or *Time*.

References

1. The National Security Board has recently relaxed restsrictions on information regarding plentisillin, known previously under the code name "Plentisillin," and has authorized release of this survey of the work done on this remarkable new drug.
2. Nadir, N. *J. Carcinogenic Chemicals* 1946, **5,** 5732.
3. A high turnover of personnel was one of the problems encountered at this step.
4. Nadir, N., Die Argentine Berichte, 1940, **10,** 452.
5. Private communication.
6. The krypton was determined bacteriologically with "kryptonless" Neurospora, a recently developed mutant.

7. Privy communication.
8. Bloshemberg, H., NY Journal-American, 1945, **92,** 42.

Norman Radin
Preprint distributed at the AAAS meeting
27–30 March, 1946, with a spoof *Science* cover

Some Observations on the Diseases of *Brunus edwardii* (Species nova)

Summary

The correct specific and generic terminology for *Brunus edwardii* is discussed and the results given of a survey involving 1599 complete specimens and 539 miscellaneous appendages. These results indicate that primary infectious agents do not occur, and that the species is safe for children to handle. Suggestions as to the future role of the profession in relation to this species are made.

Introduction

For more than a century, this species has been commonly kept in homes in the United Kingdom and other countries in Europe and North America. Although there have been numerous publications concerning the behavior of individuals (Milne, *The House at Pooh Corner*, etc.) there have been no serious scientific contributions, and a careful search of the literature, using abstracting journals and computerized data retrieval systems, have failed to reveal any comprehensive survey of the diseases of these creatures. A few of these previous publications include references to certain disease syndromes, and Milne refers to obesity associated with the excessive intake of honey, and to psychological disturbances associated with territorial disputes with Tiggers, Heffalumps, and even small children. One publication concerning a certain individual known as Paddington, refers to the animal receiving teatment from medical practitioners without a veterinary qualification. These records emphasize two disturbing factors, firstly, the obvious need for treatment of diseased individuals, and secondly the infringement of the Veterinary Surgeon's Act of 1966 that would presumably be involved if such animals were treated by any person not on the Veterinary Register.

The British Veterinary Association has many nonterritorial divisions, and it would have been logical to assume that either the British Small Animal Veterinary Association, or the British Veterinary Zoological Society would have shown an interest in this species. It is unfortunate that apparently none of the Officers of these associations had either the mentality or the foresight of the present authors to instigate a study similar to that published in this paper.

Pet ownership surveys have shown that 63.8 percent of households are inhabited by one or more of these animals, and there is a statistically significant relationship between their population and the number of children in a household. The public health implications of this fact are obvious, and it is imperative that more be known about their diseases, particularly zoonoses or other conditions that might be associated with their close contact with man. The results of this survey may also be relevant to the recent Rabies (Importation of Mammals) Order (1971), and could necessitate possible changes in the quarantine regulations.

The importance of avoiding the use of colloquial names in scientific contributions has been stressed by Keymer et al. (1969), but previous publications have apparently only used the term "teddy bear." Preliminary studies have suggested that this term might include several different strains, if not species. However, it was found that teddy bears will accept cutaneous, and even limb grafts from other bears without showing any signs of rejection. These findings obviously indicate that all teddy bears are genetically homozygous and of the same species. We therefore consider the correct generic and specific terminology to be *Brunus edwardii.*

Materials and Methods

Source of Material for Survey

A total of 1598 specimens of *Brunus edwardii* was examined. Of the 1600 owners approached, 1599 agreed to examinations of their bears, and the majority were able to provide a comprehensive case history. One specimen was eventually unavailable, as it was in quarantine because the owner was affected with rubella.

A further 539 miscellaneous appendages were made available for examination by nurseries, schools, and children's hospitals in the London area. These specimens were in a delapidated condition, but careful grafting restored 136 intact bears, with only one surplus ear, which has been stored in liquid nitrogen for future use. No case histories were available for this latter group.

Examination Techniques

Examinations were carried out as quickly as possible, because many owners were reluctant to be parted from their bears for long. No restraint was necessary, as the bears showed no apprehension and were obviously used to being handled. An attempt was made to record body temperature, but this was abandoned, as all specimens appeared to be homoisothermic. Each bear was given a thorough external examination, and data were collected on approximate age, weight, condition and color of coat, and physical disabilities. Stuffing condition was assessed by careful palpitation. Where necessary, radiographs were taken, and biopsies obtained to identify the stuffing material. Subcutaneous and deeper tissues often protruded from superficial abrasions and, where necessary, a small seam incision was made, a sample taken, and the opening sutured with Coates Machine twist 30, using a standard Millwards darning needle. Voice boxes, where present, were tested by percussion and auscultation.

The psychological state of the bear was assessed by examining the facial expression, and also by investigating the case history with special reference to the frequency and duration of association with children.

Results of the Survey

One thousand five hundred and ninety-eight animals, plus miscellaneous appendages were examined. Classification of the findings was attempted, but almost all cases were of multifactorial etiology, and it was impossible to determine the primary agent. Similar lesions occurred at many different sites, making systematic tabulation of results impracticable. No primary pathogens were isolated, and the predominant cause of pathological change was external mechanical trauma, which was either severe and sudden in onset, causing loss of limbs and appendages, or more insidious giving rise to chronic wear and tear.

Commonly found syndromes included coagulation and clumping of stuffing, resulting in conditions similar to those described as bumblefoot and ventral rupture in the pig and cow respectively, alopecia, and ocular conditions that varied from mild squint to intermittent nystagmus and luxation of the eyeball. Micropthalmus and macropthalmus were frequently recorded in animals that had received unsuitable ocular prostheses.

Ninety-eight percent of the cases examined were jaundiced, but since no samples could be obtained for the Van den Bergh tests, the reason for this condition remained obscure, and it was concluded that icterus was probably a normal state for this species.

The following case notes illustrate the complexity of both the causes and resulting manifestations of disease in the species.

Case 1: A six-month-old bear, owned by a four-year-old male, was found to be suffering from acute dyslalia, torticollis, and loss of one lower limb. The general condition of the animal was good, with a normal thick pelage. The injury had been the result of disputed ownership. Treatment of the torticollis by manipulative therapy, and surgical replacement of the limb, were uncomplicated. The dumbness was the result of a ruptured acoustic membrane, and complete renewal of the voice box was necessary. This involved laparotomy, removal of the damaged organ from its surrounding viscera, and careful positioning of the replacement so that the acoustic membrane faced ventrally to prevent the development of muffled speech.

Case 2: A young bear owned by a child of six months was found to be suffering from "soggy ear" when removed from the owners's cot one morning. Edema of the pinna was a commonly recurring condition in bears belonging to children under 18 months, who slept with an ear firmly clamped in their mouths. Treatment consisted in removal from the owner, lavage, and drying in an airing cupboard.

Case 3: A ten-year-old bear, that had been owned succes-
sively by three siblings. The normal yellow coat color had
changed to a dirty grey, there was extensive alopecia that had
progressed to "threadbearness" over the ears, nose, and limb
extremities. The axillary and inguinal seams were weak, resulting in intermittent dislocation of the limbs, but there was no
herniation of stuffing. Old age, and persistent handling with
transport by one limb were the main reasons for the chronic
debility, for which there is no satisfactory treatment.

Case 4: A six-year-old bear belonging to an aspirant
nurse. The animal had a distinctive cheesy odor and there was
a dried white oral and nasal discharge on the facial fur. This
was the only case where a baceriological examination resulted
in the isolation of an organism, *Lactobacillus acidophilus*,
which was obviously an opportunist invader of the contaminated coat. There was no evidence of digestive disturbance.
The owner was encouraged to treat the animal herself by wash-
ing it thoroughly.

Case 5: A 16-year-old bear, with an asymmetrical
expresssion and obvious emotional disturbance, found at the
back of a cupboard. After the removal of superficial dust, the
coat condition was seen to be good, but the animal had a permanent squint, due to careless replacement of the right eye with a
shoe button. Tracing of the case history revealed that this bear
had suffered recurrent unilateral ocular prolapse that had progressed to total rupture of the filamentous orbital attachments,
and loss of the eye. It was hoped that a new owner might be
found for this animal, and that with a newly matched pair of
eyes his expression and psychological state might improve.

Case 6: An aged, cobweb-covered bear, found in an attic.
Its general condition was poor, with loss of a forelimb and
herniation of stuffing. The frontal seam was ruptured, expos-
ing a rusted voice box with helical weakness. The animal was
heavily infested with commensals, which included a pair of
Mus musculus with two generations of young, a total of 23
individuals. Specimens of *Laelaps hilaris, Leptosylla segnis,
Nosopsyllus, fasciatus, and Lenisma saccharina* were found in

the right inguinal region, and the border of the left pinna was being eroded by clothes moths *(Tineola bissilliella)*. Treatment of this case included vigorous shaking, dusting with pyrethrum, a stuffing transfusion, and a forelimb graft.

Discussion

This survey has revealed many facts of interest to both the comparative pathologist and the clinician. It is with considerable relief that it can be recorded that *Brunus edwardii* appears to be resistant to any pathogenic organisms and cannot, therefore, be affected by any zoonotic condition. However, this species can be involved in a variety of commensal relationships, as illustrated by Case No. 6. Future investigations, using gnotobiotic techniques, might provide further fascinating information on such associations.

Teddy bears can act as transitory mechanical vectors of human pathogens. Although superficial contamination with rubella virus has no direct effect on the bear, the unskilled treatent of carrier teddies can result in serious secondary disease. Examples found included a singed integument caused by overheating during decontamination in a domestic oven, and coat discoloration caused by treatment with an unsuitable disinfectant.

True diseases of *Brunus edwardii* can therefore be classified as either traumatic or emotional. Acute traumatic conditions, characterized by loss of appendages, are often the result of disputed ownership. Chronic traumatic conditions are usually associated with normal wear and tear, and are not necessarily deterimental, as there appears to be a statistical relationship between the presence of such lesions, the lack of emotional disturbance, and the affection given by the owner.

Emotional disturbances are either apparent or inapparent. Apparent emotional disturbances are recognized by changes in facial expression, and in almost all cases the condition is the result of unskilled remedial surgery. Inapparent emotional disturbances are not fully understood, but seem to be related to the fact that an unloved teddy is an unhappy teddy. Few adults (except perhaps the present authors) have any real affection for the species, and as children mature, their teddy bears may be ne-

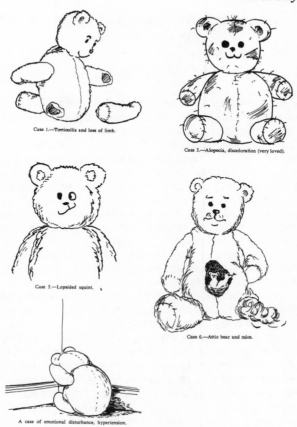

Case 1.—Torticollis and loss of limb.

Case 3.—Alopecia, discoloration (very loved).

Case 5.—Lopsided squint.

Case 6.—Attic bear and mice.

A case of emotional disturbance, hypertension.

glected and relegated to an attic or cupboard, where severe emotional disturbances develop.

The authors consider it significant that *B. edwardii* appears to be classless in both the taxonomic and socioeconomic sense.

The authors believe that their time, effort, and skill will have been completely wasted if these findings do not stimulate the practitioner to take a greater interest in the clinical problems described. It is hoped that this contribution will make the profession aware of its responsibilities, and it is suggested that veterinary students be given appropriate instruction, and the postgraduate courses be established without delay.

D. K. Blackmore, D. G. Owen, and **C. M. Young**
The Veterinary Record, Vol. 90 (14), April 1, 1972 p. 382

American Safety Code and Requirements for Dry Martinis

Foreword

The need for an American Standard Dry Martini has been widely recognized by many segments of the manufacturing, distributing, and consuming public since the martini cocktail's appearance.

Few persons alive today are likely to recall that the original martini cocktail was a mixture of equal parts of sweet or Italian vermouth and 80- to 86-proof dry or English gin. It was from the popular brand of sweet vermouth (Martini & Rossi) that the drink derived its name. Curiously, when dry or French vermouth was substituted for sweet vermouth, the drink was still called a dry martini cocktail, although it was not until many decades later that the Italian firm of Martini & Rossi began to bottle an imitation French vermouth.

At various times in the development of the classic or American Standard dry martini, consumers have exerted strong influence to improve the character of this tonic by asserting rigid preferences in the the proportion of gin to vermouth in the final product. However, the manufacturing and distributing elements of the martini picture have chosen to adhere to a standard based on color rather than proportion.

As a result, consumers called for drier and drier martinis, by which they meant ones prepared with less and less vermouth, and distributors responded by providing a paler-looking drink with either the same proportion of ingredients or else more vermouth of a water-white color. It became evident that some area of meeting was necessary if the venerable dry martini was to remain alive.

It is hoped that the pioneering work of the committee will be well received by all elements of the industry and that when the committee sobers up it will be in condition to consider further developments in the state of the art.

The ASA Sectional Committee on Liquids Management, K-100, which reviewed and approved this standard, had the following personnel at the time of approval:

Gilbey Gordon Booth, *Chairman*
Bertram Stanleigh, *Secretary*

Organization Represented	*Name of Representative*
American Society of Bar Supporters	W. C. Fields
Association of Amiable Alcoholics	Lyle Sachs
Boilermaker Dispensers and	Burt Piel
Consumers of America	Harry Piel (Alternate)

Organization Represented	*Name of Represenyative*
Business Decisions Coordinating Committee	Nielson Trend
Cocktail and Aperitif Imbibers Council	J. Noilly Prat
Dry Wine and Cocktail Institute	Floribunda Trellis
Executive Planning Council	Trevelyn Trevelyn
Gin Council of America	Horace Beefeater
Housewarmers Association	Merryweather Bellweather
Intemperates Anonymous	*Anonymous*
Legislative Lapses Codifying Commission	Hon. Bobby Baker
Manufacturers Practices Association	Grafton P. Underhand
National Council of Developers and Coordinators	Junius Q. Underfoot
Olive Institute	Terrence Pitt
Political Divisions Coordination Council	Harry Underarms
Public Relations Institute	Ubiquitous Legrand
Standard Stirrers of the United States	Boylan Murdock
Traditions Maintenance Society	Rev. Coates Plymouth
Vermouth Council	Martin Rossi
Water Conservation League	Noah Drysdale
Xmas Comes But Once a Year	C. Kringle
Yankee Know-How Institute	G. O. Home
Zymurgy Standards Commission	Noah Webster

The technical Committee for American Standard Safety Code and Requirements for Dry Martinis, K100.1, consisted of:

Archie Allen, Jr.	Maril Pisciotta	Norris Smith
Joan Nestle	Kim Sinclair	Sidney Taylor

Scope

This standard on dry martini cocktails includes nomenclature, sizes, ingredients, proportions, mixing methods, and test procedures. It applies to martini cocktails prepared for personal consumption, for distribution in bars, restaurants, and other places of public gathering, and to cocktails served in the homes or offices of business and social acquaintances.

1. Definitions

Dry Martini. A cocktail made with English or American dry gin of at least 86 proof and dry vermouth, preferably French in origin, in accordance with requirements of this American Standard.

Extra Dry Martini. A meaningless expression used loosely by waiters and bartenders. It is frequently the excuse for a supplementary charge and is often characterized by the inclusion of excessive melted ice or an abundance of water-white vermouth.

French Vermouth. A term generally applied to dry varieties of vermouth whether they are actually produced in France or in some other country. The true French product is an infusion of herbal extracts in a undistinguished white wine of the Midi region. Characteristically pale brown in color, it has recently been produced in much lighter versions.

Gibson. An unpardonable form of perversion. *See* Onion soup.

Gin. An infusion of juniper berries and other extracts in grain alcohol. While the drink is generally credited to be Dutch in origin, the variety that evolved in England during the eighteenth and nineteenth centuries is the basic type employed in all American Standard dry martinis. Currently, a barely acceptable product is distilled in the United States, but it seldom aspires to more than minimal requirements.

Italian Vermouth. The sweet aperitif wine that was originally combined with gin to produce the martini cocktail. It is an ingredient in many drinks, but it is never employed in the preparation of dry martinis.

Lemonade. A term applied to drinks which have been subjected to the peel of a lemon. There is no place for the rind of any citrus fruit, or its oils, in an American Standard dry martini.

London Dry Gin. A term encountered on the labels of imitation English gins. Many of these specimens are moderately palatable and approach the minimal levels of the American Standard.

Martini. A broad term that can frequently lead to differences of opinion, but which will invariably lead to a state of inebriation. Originally the first name of a firm of wine merchants, it can mean anything from a glass of sweet vermouth (in the British Isles and on the Continent), to a martini cocktail, or an American Standard dry martini.

Olive. The fruit of a Spanish tree, the olive is encountered in its green state, pitted and unstuffed, in the classic dry martini. While olives are normally considered superior if their size is great, when included in a dry martini, the small, or cocktail variety, is mandatory. A list of maximum displacements for olives in American Standard dry martinis is shown in Table 1. The absence of an olive is not critical provided there is no diminution of the total volume of the drink.

Onion Soup. The unholy abomination produced by the introduction of one or more pickled onions into a dry martini cocktail.

Rocks. The solid state of H_2O on which an American Standard dry marini is never served.

Vodka. A distilled alcoholic beverage made originally from potatoes, but now encountered in grain alcohol versions. It may be clean, palatable, and nonlethal, and when encountered in this form, is a fitting accompaniment for fresh cavier. It is never employed in a dry martini.

2. Sizes

2.1. Basic Nomenclature. The American Standard dry martini shall come in the following three sizes:

(1) Regular   not less than 3-1/2 ounces
(2) Large   not less than 5 ounces
(3) Double   not less than 7 ounces[1]

2.2. Recomended Ingredients. Although the subject of ingredients is more fully covered in Section 3, Ingredients, it is advisable to make the following observations concerning ingre-

---

[1]Not more than one olive shall be used in the double dry martini, and its size shall be no larger than that shown in Table 1.

dients in relation to the size of the drink. In the regular dry martini, it is recommended that no gin of less than 90 proof be employed, with strengths of 94.4 and 100 proof preferred. In a large dry martini, 90 proof gin is the preferable variety, and in the double dry martini it is considered unwise to use any gin of greater than 90 proof. In the double dry martini, those who protest against the rising tide of conformity may even be justified in the employment of 86 proof gin (*see* Table 2).

3. Ingredients

3.1. General. Only the following three ingredients shall be used in the preparation of any American Standard dry martini:

(1) Gin
(2) Dry Vermouth
(3) Olives

3.2. Gin. As the chief ingredient of the American Standard dry martini, it is essential that the gin employed shall conform to the highest standards in color, flavor, aroma, and alcoholic content.

3.3.1. Color. The color shall be either water-white or faintly blue. No pale yellow tints or slightly grey tinges shall be acceptable.

3.2.2. Flavor. The flavor shall be full, clean, and lacking in harshness. When rolled on the tongue and sucked through the teeth, the fluid shall exhibit a soft, supple quality without any trace of oiliness.

3.2.2.1. Aftertaste. Following the swallowing of the gin there shall remain in the mouth for a period of no less than 30 seconds an agreeable sensation vaguely reminiscent of the full flavor of the gin. There should be no heightened intensity to the character of any one flavor element within the gin.

3.2.3. Aroma. The smell shall be delicately assertive, combining the aromatic elements of the essence of juniper berries and pure grain alcohol.

3.2.4. Alcoholic Content. Any of the following commercial strengths of gin shall be acceptable, with the exception noted:

(1) 86 proof[2]   (5) 96 proof
(2) 90 proof   (6) 100 proof
(3) 94 proof   (7) Any gin that exceeds 100 proof
(4) 94.4 proof[3]

3.3. Vermouth. Dry, often called French, vermouth shall be of excellent taste, exhibiting no tendency toward sweetness, acidity, or courseness. It shall be free of deposit and possess a delicate, fragrant aroma.

3.3.1. Use of Vermouth. The employment of vermouth in an American Standard dry martini shall not be mandatory, provided no other ingredient is employed as a substitute.

3.4. Olives. Although the use of olives is not encouraged, nothing in this specification shall be construed to mean that the inclusion of an olive will not be acceptable, provided it conforms to Table 1 and sub-paragraphs 3.4.1 and 3.4.2

3.4.1. Color. The olive shall be in an unripe state and of the color known commercially as olive green. The color shall be uniform, without brownish spots or patches of more intense pigmentation.

3.4.2. Contents. While green olives are acceptable for many purposes in an unpitted state or with their pits replaced by such exotic items as pimento, almonds, anchovies, etc., for employment in the preparation of the American Standard dry martini, only olives without pits or stuffing shall be used.

4. Proportions

No other single element is more critical in the preparation of the American Standard dry martini than the proportion of gin to dry vermouth. In specifying proportions it is necessary to take into account both the size of the drink and the strength of

[2]In this strength, only Booth's "House of Lords" gin meets minimal standard requirements.

[3]For practical purposes, this strength may be considered jointly with 94 proof.

TABLE 1
Maximum Permissable Olive Displacement

Maximum olive size, in^3	Glass capacity, ounces[5]	Normal size
.4730	3-1/2 (– 0 + 1/2)	Regular
.5221	5 (– 0 + 1)	Large
.473[4]	7 (– 0 + 8)	Double

the gin. Table 2 indicates maximum quantities of vermouth that shall be acceptable. There are no minimum requirements for vermouth in an American Standard dry martini.

5. Mixing Method

5.1. Apparatus. A container, stirrer, calibrated measure, ice strainer, and 60 watt incandescent lamp are all items that may be required in the preparation of a dry martini. Their necessity will be determined by the mixing method to be employed.

5.2. Methods. Dry martinis may be mixed in any one of the following manners:

5.2.1. Stirring Over Rocks. In this method, proper proportions of the stipulated ingredients are poured into a container over solid pieces of ice. Crushed or cracked ice shall not be used, and at least 90 percent of the ice employed shall be in pieces at least 1 cubic inch in size. Following an interval of not less than 30 seconds are not more than 1 minute, the ingredients shall be stirred by one of the methods indicated in Fig. 1. Stirring shall be vigorous enough to encourage a blending of the gin and vermouth, but gentle enough to ensure the slightest amount of melted ice.

[4]*See* footnote 1.
[5]All measurements are to be made with glasses filled to the brim.

TABLE 2
Proportions

Gin proof	Minimum parts gin	Maximum parts vermouth	Nominal drink size
86	20	1	Double
90	17	1	Double or Large
94	17	1	Large or Regular
96	17	1	Regular
100	16	1	Regular

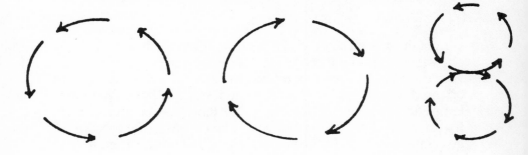

Counterclockwise   Clockwise   Both

Fig. 1. Stirring Patterns.

   5.2.2. Blending of Refrigerated Ingredients. In this method, both the gin and vermouth are refrigerated to a temperature no higher than 32° Farenheit and then mixed without the addition of ice. Care shall be taken to refrigerate the mixing container as well as the ingredients.
   5.2.3. Radiation. This method produces martinis of the proper degree of dryness with an accuracy not even approached by the preceding methods. It also makes it possible to produce and store proper martini cocktails by the bottleful. As indicated in Fig. 2, a 60-watt incandescent lamp is placed on a flat surface exactly 9 inches from a sealed bottle of vermouth. A sealed bottle of gin is placed on the other side of the bottle of

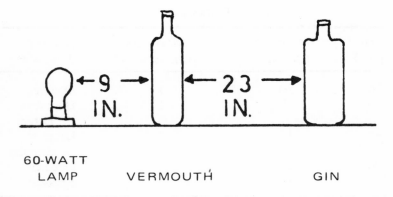

60-WATT
LAMP   VERMOUTH   GIN

Fig. 2. Radiation Mixing Method.

vermouth at a distance of 23 inches. Care shall be taken to align the bottles so that the rays of the lamp pass through the vermouth bottle directly into the gin bottle. Labels shall be so oriented that they do not hinder such passage of light. With the lamp and bottles suitably arranged, the lamp may be illuminated for an interval of 7 to 16 seconds. The duration of exposure is governed by the color of the bottles. Clear bottles require the shortest exposure, dark green bottles demand the longest exposure.[6]

6. Test Methods

The testing of the American Standard dry martini requires a high degree of skill, experience, and selfless dedication. No known scientific apparatus has yet been developed that can match the sensitivity of the palate of a qualified American Standard dry martini taster. In testing, the taster shall watch for lightness of color, absence of sediment, a delicate aroma that effectively combines the scent of the juniper with the herbal infusion in the vermouth, a taste that is both sharp and clean with a faint body, and a light delicate aftertaste.

[6]Subcommittee 20-1 is presently engaged in the development of standards for uniform optical density of gin and vermouth bottles.

Life Among the Econ

The Econ tribe occupies a vast territory in the far North. Their young are brought up to feel contempt for the softer living in the warmer lands of their neighbors, such as the Polscis and the Sociogs.

Caste and Status

The Econ word for caste is "field." Caste is extremely important to the self-image of the Econ, and the adult male meeting a stranger with the phrase "Such-and-such is my field." The basic social unit is the village, or "dept." The depts of the Econ always comprise members of several "fields."

The dominant feature, which makes status relations among the Econ of unique interest to the serious student, is the way that status is tied to the manufacturing of certain types of implements, called "modls." The status of the adult male is determined by his skill at making the "modl" of his "field." The facts (a) that the Econ are highly status-motivated, (b) that status is only to be achieved by making modls, and (c) that most of these modls seem to be of little or no practical use, probably account for the backwardness and abject cultural poverty of the tribe.

Grads, Adults, and Elders

The young Econ, or "grad," is not admitted to adulthood until said Econ has made a modl exhibiting a degree of workmanship acceptable to the elders of the dept in which the apprenticeship is being served. Adulthood is conferred in an intricate ceremony, the particulars of which vary from village to village. In the more important villages, furthermore, the young adult must continue to demonstrate an ability at manufacturing these artifacts. If the Econ fails to do so, the dear student is turned out of the dept to perish in the wilderness.

If life is hard on the young, the Econ show their compassion in the way that they take care of the elderly. Once elected an elder, the member need do nothing and will   still be well taken care of.

Totems and Social Structure

Modl has evolved into an abstract concept that dominates the Econ's perception of virtually all social relationships. Thus, in explaining to a stranger, for example, why the Econ holds the Sociogs and the Polscis in such low regard, friend Econ will say that "they do not make modls" and leave it at that. There is considerable support for those who refer to the basic modl as the *totem* of the caste.

Myths and Modls

The Math-Econ are in many ways the most fascinating, and certainly the most colorful, of Econ castes. There is today consideraable uncertainty whether the "priest" label is really appropriate for this caste, but it is as least easy to understand why the early travelers came to regard them in this way.

The Math-Econ make exquisite modls finely carved from bones of walrus. Specimens made by their best masters are judged unequalled in both workmanship and raw material by a unanimous Econographic opinion. If some of these are "useful", it is clear that this is purely coincidental in the motivation for their manufacture.

There has been a great deal of debate in recent years over whether certain Econ modls and the associate belief-systems are best to be regarded as religious, folklore and mythology, philosophical and "scientific," or as a sports and games. The cermonial use of modls and the richness or the general Econ culture in rituals has long been taken as evidence for the religious interpretation. But, as one commentator puts it, "If these beliefs are religious, it is a religion seemingly without faith."

The sports and games interpretation has gained a certain currency due to accounts of the modl ceremonies of the Intern caste. But even here it is found that, though the ceremony has the outward manifestations of a game, it has to the participants something of the character of a morality play that in essential respects shapes their basic perception of the world.

The Econ and the Future

The prospect for the Econ is bleak. Their social structure and culture should be studied now before it is gone forever. Even a superficial account of their immediate and most pressing problems reads like a veritable catalog of the woes of primitive peoples in the present day. They are poor—except for a tiny minority, miserably poor. In the midst of their troubles, the Econ remain as of old a proud and warlike race.

But they seem entirely incapable of "creative response" to their problems.

Econ political organization is weakening. The basic political unit remains the dept and the political power in the dept is lodged in the council of elders. The foundation of this power of the elders has been eroding for some time, however. Respect for one's elders is no more the fashion among the young Econ than among young people anywhere else.

The Econ adult used to regard himself as a life-long member of his dept. This is no longer true—migration between depts is nowadays exceedingly common.

Under circumstances such as these, we expect alienation, disorientation, and a general loss of spiritual values. And this is what we find. A typical phenomenon indicative of the break-up of a culture is the loss of a sense of history and growing disrespect for tradition. Having lost their past the Econ are without confidence in the present and without purpose and direction for the future.

Some Econographers disagree with the bleak picture of cultural disintegration just given, pointing to the present as the greatest age of Econ Art. But it is not unusual to find some particular art form flowering in the midst of the decay of a culture. It may be that such decay of society induces this kind of cultural "displacement activity" among the talented members who despair of coping with the decline of their civilization. The present burst of sophisticated modl-carving among the Econ should probably be regarded in this light.

Axel Leijonhufvud
condensed from *Western Economic Journal*
Vol. 11, September 1973, pp. 327–337

Spoof

The Art of Making Saltpetre

During the American Civil War the Confederates, to provide nitrate for making gunpowder, had to resort to all sorts of devices, such as digging out and leaching the earth from old smokehouses, barns, and caves and making artifical beds of all sorts of nitrogenous matter, and they had agents for the purpose in every town and city.

The officer at Selma, Alabama, was particularly energetic and enthusiastic in his work and put the following advertisement in the Selma newspaper on October 1, 1863:

"The ladies of Selma are respectfully requested to preserve the chamber lye collected about the premises for the purpose of making nitre. A barrel will be sent around daily to collect. John Haralson, Agent, Nitre and Mining Bureau, CSA."

This attracted the attention of army poets and the first of the following two effusions resulted. It was copied and privately circulated across the Confederacy. Finally it crossed the line, and an unknown Federal poet added the "Yankee's View of It."

Confederate View of It

John Haralson, John Haralson—you are a wretched creature;
You've added to this bloody war a new and useful feature.
You'd have us think, while every man is bound to be a fighter
The Ladies, bless the pretty dears, should save their pee for nitre

John Haralson, John Haralson, where did you get the notion
To send the barrel 'round to gather up the lotion?
We thought the girls had work enough in making shirts and kissing,
But you have put the pretty dears to Patriotic Pissing.

John Haralson, John Haralson, pray do invent a neater
And somewhat less immodest way of making your saltpetre
For 'tis an awful idea, John, gunpowdery and cranky
That when a lady lifts her skirts, she's killing off a Yankee.

Yankee View of It

John Haralson, John Haralson, we've read in song and story
How women's tears, through all the years, have moistened fields
of glory—
But never was it told before, how 'mid such scenes of slaughter
Your Southern beauties dried their tears and went to making water.

No wonder that your boys were brave; who couldn't be a fighter
If every time he fired his gun, he used his sweetheart's nitre?
And vice versa, what could make a Yankee soldier sadder
Than dodging bullets fired by a pretty woman's bladder?

They say there was a subtle smell that lingered in the powder
And as the smoke grew thicker and the din of the battle louder
That there was found to this compound one serious objection—
No soldier boy could sniff it without having an erection!!!

Richard C. Sheridan
CHEMTECH
April, 1976

Locke's 'Life on the Moon'

One of the most difficult feats to achieve in science is a successful hoax. It requires experience and talent to deceive a large number of normally sceptical people into believing in a bogus discovery, and to maintain the deception.

The Piltdown hoaxer, whoever the culprit may have been, plainly had both these qualities, constructing a relic of early man that convinced professional scientists for 41 years. So also did the New York journalist Richard Adam Locke, perpetrator in 1835 of the Great Moon hoax.

Locke's achievement was to persuade hundreds of thousands of readers of the New York Sun that responsible scientists had found an intelligent civilization of winged men living on the moon.

He set about this fraud in a most businesslike fashion. Day after day, he told astonished newspaper readers that these discoveries were being made by the great astronomer, Sir John Herschel, in his newly installed observatory a few miles from Cape Town.

It was true that Herschel had an observatory near Cape Town, but nearly everything else that Locke said about him was false. Yet it sounded authentic. Locke knew enough about astronomy to write convincingly about Herschel's telescope, although in a wildly exaggerated fashion, making it ten times larger and thousands of times more powerful than it actually was.

Locke had the perfect manner that a successful conman needs to deceive both his editor and his readers. Even in appearance, he was more like the popular image of a scientist than a newspaper reporter. His face was high-domed and intellectual, his dress meticulous, his manner grave and professional.

He made no claim to have interviewed Herschel personally; much more cleverly, he stated that he was quoting Herschel's reports in the *Edinburgh Journal of Science* sent to him, allegedly, by a devoted friend of Herschel's in Scotland, a certain Dr. Andrew Grant.

Might not so elaborate a story contain some truth? It contained none: but for a time, Locke got away with it.

In a world without aircraft or radio, there was no one to point out that the Edinburgh journal to which Herschel had in fact often contributed had ceased publication two years before, and that 'Dr. Grant,' who was thanked so politely in Locke's newspaper, did not exist.

It was two months before Herschel learned of his own fabulous lunar 'discoveries,' by which time Locke himself had disappeared.

Locke wrote with a kind of high-sounding pseudotechnical phraseology that today sounds like dialog from *Star Trek*. To explain how Sir John Herschel was able to see these extraordinary creatures jumping around on the moon (a feat that would be impossible even through modern telescopes), he described an imaginary conversation between Herschel and his real-life friend, the optical inventor Sir David Brewster.

"After a few minutes' silent thought, Sir John diffidently inquired whether it would not be possible to effect a transfusion of artificial light through the focal object of vision..."

In plain language, this could only mean one thing; to erect a giant searchlight with which to see the lunar creatures more clearly. But of course Locke could not afford to use plain language. To be convincing, his explanation had to be unintelligible.

Brewster agrees that this 'transfusion of artificial light' would be possible, and the two of them set to work. Soon the advanced life-forms on the moon became apparent.

They are shaped like a humans—about four feet tall, but with wings. Their faces *"of yellowish flesh color, are open and intelligent in their expression, having a great expansion of forehead."*

These beings talked, bathed in lakes, shook their wings, and flew. Herschel is made to explain that *"they are doubtless innocent and happy creatures, notwithtanding that some of their amusement would but ill comport with our terrestrial notions of decorum."*

The beginning of Locke's downfall came when he tried to be too clever. He concluded his articles with fatal reference to *"forty pages of illustrative notes, which we omit as being too mathematical for popular comprehension."*

American astronomers had long been suspicious, and now they pounced. A scientific deputation arrived in the offices of the New York Sun. Locke tried to charm them with protestations of being flattered at meeting men of such distinction, but to no avail. Where, they demanded, were those mathematical notes?

Locke feigned embarrassment. The timing of their request was really most unfortunate, for the notes were at the printers.

Which printers, they asked?

Locke gave them an address, and they set off at once. But Locke himself traveled there more swiftly, to instruct his friend at this printing works to tell the scientists, when they arrived, that the notes had been sent elsewhere.

The investigators had a fruitless quest. In the words of the scientific historian Willliam H. Barton: "Locke sent them on a wildgoose chase from one printer to another, all the while shortcutting through lanes and alleys to tell his friends where to direct them next. He was no impractical genius."

When shall we see the next great scientific hoax? My hopes rose in 1978 when the Cutty Sark Scotch Whisky Company offered a prize of £1 million to anyone who could discover an alien spacecraft on earth. With the promised £1 million to spend, surely somebody could fake up some plausible-looking contraption and give the company a well-earned fleecing? But sadly it has not happened.

Yes, sadly is the right word. Scientific hoaxes are not necessarily a bad thing. They keep honest scientists on their toes.

Adrian Berry
High Skies and Yellow Rain
London, Daily Telegraph, 1983

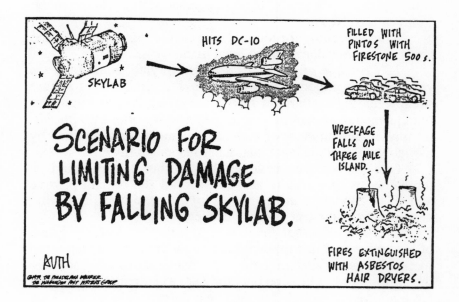

Scenario For Limiting Damage
by Falling Skylab

Can Skylab Lawyers Fool Mother Nature?

This is the week the 85-ton space station, Skylab, plagued by a diminishing orbit, will strike Earth's denser atmosphere, skipping along its upper reaches like a silver dollar skimmed across the Potomac, eventually to lose the momentum of its flight and plunge.

Smaller parts will be consumed in their fiery reentry, but some larger parts, though scorched, will rush through the atmosphere back toward the ground whence they came. It is not unlikely that pieces as large as two tons will reach Earth, and it is hoped that these will not fall on populated areas.

After failing at various technological attempts to keep Skylab in orbit and intact until it could be reached by a rescue crew in the space shuttle, NASA engineers have abandoned it to the forces of nature. The only way a potentially disastrous reentry can now be prevented is if the NASA legal department can come up with a successful defense against Mother Nature in the Court of Last Resort.

After searching for loopholes in current physical laws, the lawyers are considering these legal maneuvers:

1. Principle of In Loco Parentis. Skylab, being only six years old, is a minor in the eyes of the law, and Mother Nature became its legal guardian ipso facto when NASA engineers relinquished their control. Thus, the principle of *in loco parentis* applies, and any acts of violence or vandalism committed by Skylab would be the legal guardian's responsibility.

One lawyer even believes that Mother Nature could be threatened with a child-abuse suit should she remain passive and allow Skylab to inflict bodily injury upon itself by falling headlong from the cradle of the heavens.

2. Environmental Protection Standards. Another group of lawyers would like to force Mother Nature to file an Environmental Impact Statement before her charge could fall back into the atmosphere. The time it would take to prepare, defend, and have approved such a statement would certainly give NASA engineers time to launch the space shuttle on a mission to kidnap Skylab and deprogram it. Although NASA lawyers have been told they won't be able to fool Mother Nature, they feel it is worth a trial.

3. Unlawful Disassembly. One legal team my seek an injunction against Skylab's reentry. They argue that the jurisdiction of the Universal Law of Gravitation, by which all matter in the universe is said to be mutally attracted, has not yet been properly clarified among or ratified by the member planets of our solar system. The lawyers contend that, though interplanetary space is indeed subject to the law, planet Earth maintains legal privilege within its own territorial atmosphere. Should she choose, Earth could forbid the renegade satellite's entering her 60-mile limit, within which foreign objects can be fired upon for unlawful disassembly.

4. Newton's Loopholes. One of the most promising defenses against Skylab appears to be that being prepared by a young team of physical lawyers looking for loopholes in Newton's Laws of Motion.

Newton's First Law states that a body at rest tends to remain at rest and a body in motion tends to remain in motion in a straight line unless acted upon by an external force. The NASA lawyers argue that the gravitational attraction of Earth on Skylab is not in fact an external force and Skylab should continue out into space in a straight line tangent to its present orbit. Central to the lawyers' argument is the fact that Skylab was constructed of earthly materials and thus is an extension of the planet. If this point is to be admitted, any forces between the planet and its natural offspring cannot be considered external. Ergo, Skylab is free of external forces, save those of the other planets, and it should tend toward those other sources of attraction.

Newton's Third Law states that every action is accompanied by an equal and opposite reaction, and NASA lawyers cite the case of *Immovable Object v. Irresistible Force* to support their argument here. Since Earth does not want Skylab to strike it, the planet has the right to exercise an Immovable objection. An objection is in effect a repulsion, and therefore Earth in fact exerts a negative force on Skylab. This negative force, should it be sufficiently strong, should overcome gravity and send Skylab into an ever-higher orbit.

The NASA lawyers are not opposed to an out-of-court settlement, but because of judicial backlogs the most promising compromise could not be reached in time: having Skylab enter the atmosphere on the evening of the Fourth of July, giving the whole nation a spectacular fireworks display. With everyone watching the skies, all would have had a fair chance of getting out of the way in time.

Henry Petroski
Chicago Sun-Times

Man's Best Friend

My dog is an avid reader of *Science* (in fact he was trained on it), and after reading the letters on dog control (1 Nov. 1974, p. 394), he had this to say:

"The points raised, raised my hackles and gave me a dry nose. They go from canine to asinine. I am surprised that so much hate for humankinds' best friend could be unleashed. My family and I are blamed for the destruction of the environment, the inflation of tree-planting costs, and—as a solution—it is proposed that we be eaten. What a way to solve problems —eat the scapegoats. Miller says that many animals fill dual roles as pet and as food. I can only reply that I have never yet seen a chicken fetch slippers. He says that several cultures eat dog. On this argument I could add that several cultures have also been known to eat humans. Let the unthinkable remain uneatable. Anderson seriously bemoans the cost of dog urination problems, but until he can show that dogs kill trees (as opposed to smog, insectides, and so forth) he doesn't have a leg to stand on. Clearly he is barking up the wrong tree. Finally, Sears claims coyotes get the blame for damage done by dogs. Anyone with a nose for science knows that *Science* usually requires evidence. Sears offers no evidence for this claim. Since dogs have been guard, guide, entertainer, and companion to humans for longer than memory, I can only conclude that these writers were engaged in the human pursuit of dealing with problems by bitching."

That is what he said.

M. Ian Phillips
Science
January 17, 1975; p. 113

Maxwell Demons

In examining familiar observations in terms of molecular theory, Maxwell invented a microscopic, intelligent demon. Then he inquired what might happen in a container of gas divided into two parts by a transparent partition in which there was a tiny hole, covered by a sliding shutter. The demon was to be stationed at the hole and instructed to move the shutter to allow only fast-moving molecules to enter chamber *A* and only slow-moving molecules to enter Chamber *B*. Soon the temperature of Chamber *A*, determined by the average kinetic enery of its molecules, would rise, whereas the temperature in Chamber *B* would fall, all without the work of any outside agent (and contrary to experience). Maxwell demons have been used to clarify discussions of properties of radiation, information theory, etc. But the following quotation shows the fascination Maxwell's invention has held for just plain spoofers.

We hear the story that Maxwell, showing Kelvin some optical instrument, invited Kelvin to look through the eyepiece. Kelvin was surprised to find that, although the phenomenon described by Maxwell was undoubtedly there, so was a little human figure, an incongruity, dancing about. Kelvin could not help asking "Maxwell—but what is the little man there for?" "Have another look, Thomson," said Maxwell, "and you shall see." Lord Kelvin had another look, but was no wiser. "Tell me, Maxwell," he said impatiently, "What *is* he there for?" ... "Just for fun, Thompson!" replied Maxwell.

R. V. Jones
Bulletin of the Institute of Physics
June, 1957; p. 193

Realization of Reciprocal Networks
in Spectrometer Design

The response R of a nuclear sample to a stimulus S can be expressed, as is well known, in terms of a (complex) transfer function H, which bears the information of interest in the frequency domain:

$$R\,(\omega) = H\,(\omega)\,S\,(\omega)$$

Similar relations hold throughout linear network theory. Consider a two-part network whose transfer function in the forward direction is H and whose transfer function in the reverse direction is its inverse, H^{-1}. Such a "perfectly reflecting" network, which is termed a unitary reciprocal network, has the property that the response R obtained from forward stimulation, if reapplied to the output port, regenerates the original stimulus at the input:

$$H^{-1}R \;=\; H^{-1}HS \;=\; S$$

The simplest passive networks do not have this property; their transfer funcitons in both directions are the same. Some active networks, e.g., practical amplifiers, approach the reciprocal property somewhat more closely: A forward gain G is usually accompanied by a reverse gain approaching zero. By suitable design, it is possible to adjust this reverse gain to approach $1/G$. By doing so (and by paying suitable attention to forward and reverse group delays), it is possible to realize unitary reciprocal amplification in practice. The same is true not only of amplifiers in the usual sense, but of many other devices representable as two-port networks.

We have been engaged in the design of a new NMR spectrometer system that achieves unusual versatility in part by the incorporation of reciprocal elements to a maximal extent. Although it is not central to our design goals, such a spectrometer presents an interesting possibility: Whereas one normally presents a radiofrequency excitation at the input and obtains a spectrum at the output, one can envision applying a spectrum at the output and obtaining an *rf* "response" at the input. Though such reversibility at first sight suggests a violation of the second law of thermodynamics,* it must be kept in mind that the implied entropy reduction in the reverse direction is made possible by the work done by the power supplies of the system.

Our spectrometer operates in the pulse Fourier mode. The block diagram of that part of it in which reciprocal network characteristics have been realized or approximated is shown in the figure. The elements have been labeled according to their functions in the "Normal" forward mode.

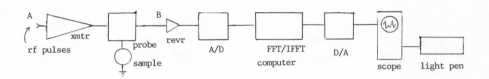

Fig. 1. Spectrometer block diagram (patents pending).

Several unique features of these elements bear mentioning in connection with the question of reciprocity:

A. The probe is based on the symmetrical traveling-wave design of Lowe, except that suitable active negative-resistance and negative-reactance elements are added in a "racetrack" configuration.

B. The main sample coil is of bifilar construction.

C. The analog–digital and digital–analog converters indicated as such are actually identical circuits incorporating heavy inverse feedback so that corresponding analog and digital signals are always present at ports regardless of which is considered as the "cause."

*Most appositely stated by Verdier (P. Verdier, *Oeuvres Scientifiques,* Éditions Beaudelairisées, Noyau-sous-Quiquette, 1869, vol. II): "Si on introduit ordure dans la derriere d'un cochon, on n'botient, qu'avec une probabilité négligéable, maïs de la bouche."

D. The Fourier transformation is realized in the software equivalent of a set of shift registers. These are arranged in a ring-counter configuration tapped at two antipodal points, at which a data set and its transform appear simultaneously (except for processing delays).

E. The light pen can serve as an input device to the oscilloscope if the operator traces the desired waveform.

It should be remarked that reciprocal behavior could easily be extended to a chart plotter at the "output." We have not yet succeeded in extending reciprocity further in the "input" direction, i.e., to pulse programmer and its controls.

In a preliminary test of these concepts, we first obtained a FT (proton) spectrum of a sample of ethanol, and photographed the scope display. An enlarged transparency was made and taped to the scope face as a guide to the operator, who then traced it with the light pen while the "input" of the spectrometer (point A in the figure) was monitored with an auxiliary oscilloscope. On each such tracing, after a short signal-processing delay, a weak rf pulse was detected at the "input." The shape of this pulse was not reproducible, presumably because of operator inaccuracies, and it was often lost in the noise.

In thinking of ways to make this effect more convincing, we realized that, because of the reciprocal characteristics mentioned, the signal at point A is of course expected to be very much weaker (by a factor equal to the transmitter forward gain) than at point B of the figure. Accordingly, the above-mentioned experiment was repeated with the monitor scope connected to B. In this case very intense pulses were observed. As a further check we returned the spectrometer to the Larmor frequency of carbon-13. When the operator traced the appropriate carbon-13 spectrum (previously recorded in the conventional mode), strong pulses were again observed at B. When the ^{1}H spectrum was traced on the scope, however, *no pulses were seen.*

In the course of these experiments, the failure of a solder connection resulted in the transmitter becoming disconnected from the probe. Of course the pulses at B were interrupted. However, we noted after some time that the level of the sample liquid in the tube appeared to have risen slightly. Continued tracing of the spectrum on the scope face verified this effect: indeed, an initially empty sample tube in the course of time accumulated a quantity of a liquid that on examination proved to have a pleasant aroma and flavor. Chemical and mass spectroscopic analyses showed that the composition changed slightly as the experiment continued, probably because of small operator errors in the tracing process, supporting the hypothesis that the sample was actually that substance whose structure corresponds to the NMR spectrum that was input.

We believe that thse findinGs have important consequences for the chemicel syntheS is. HꟿWever our present goꟿcismaximize the yiEld as R ap id llvas pOSShbl.

F. Stickney DeBouregas
(nee **John Waugh**)
NMR Conference
Asilomar, CA, 1977

The Pencil Problem—1990

When Monte Throdahl, senior vice president for Monsanto, addressed a Washington, DC, conference on toxics control, he included the following scenario. He has described it as an exaggeration to the point of outrage.

This is the story of the US pencil industry. Remember, we are looking back from our vantage point in 1990. It's strange to think that, back in 1979, just anyone could use a pencil any way one wanted to.

You see, it all started when the Occupational Safety & Health Act carcinogen policy went into effect. The graphite in the pencil leads always contained a residue of crystalline silica. And there was at least one animal test and an in vitro test indicating that crystalline silica produced tumors, so the material became regulated as a carcinogen. There was no alternative for pencils, so exposure had to come down to almost zero. Workers were put into protective clothing, and that solved the problem initially.

But then the Environmental Protection Agency, acting under the Clean Air and Clean Water Acts, which soon had their own carcinogen policies, required drastic reductions in emissions and effluents. The control technology was quite expensive, and only the largest manufacturers could afford it. This caused a flurry of antitrust suits in the early 80s when there were only three pencil makers left in the country. One of the three was split into smaller companies, but they soon went out of business since they were unable to afford increasingly stringent workplace and pollution control requirements. Then foreign pencil manufacturers began to threaten to dominate the pencil market, and our government, in an abrupt about face, allowed a merger of the two remaining companies to meet overseas competition.

The Consumer Product Safety Commission then became concerned with what the newspaper headlines were calling the "pencil problem." Rubber erasers could be chewed off and choke small children. The sharp points of pencils could also be dangerous. There were residual solvents in the paint used on pencils, and pencil-chewing seemed to be a more widespread habit than anyone had realized. Printing a legend on each pencil that said "This Pencil Could Be Hazardous to Your Health" did not seem to affect consumer pencil habits, a Harvard study indicated. In fact, the study found additionally potentially harmful uses, such as stirring coffee. This led FDA to declare that harmful sustances could be dissolved out of the pencil into the coffee, and thus pencils violated food additive laws, including the Delaney amendment.

Trying to salvage its business, the pencil company began to make pencils without paint, without erasers, and with only soft leads so they could not hold a sharp point. But consumers were outraged, and sales declined.

Then someone invented a machine that could measure crystalline silica below the parts-per-trillion level, and workplace, air emission, water effluent, and waste disposal regulations required that the best practicable technology be used to reach this new low level. The pencil company was threatened with financial ruin because of the large sums needed to purchase new control equipment. There were those that wanted to ban pencils entirely under the Toxic Substances Control Act, but the government decided that pencils were necessary, particularly since they were used to write new regulations. Besides, the Senators from the state where the pencil company was located declared that pencils were as American as baseball, and should not be replaced with ball-point pens.

So the government bailed out the pencil company with a large, guaranteed loan—called a Chrysler loan in those days. But, of course, that was only a temporary measure, and to protect the pencil business, the government eventually nationalized it.

It was comforting to know, after all, that society is being protected against a danger that was so obvious we didn't even notice it for many years. There are still those who complain about paying $17 for a pencil, but you really can't put a price tag on health or safety.

Monte Throdahl
Chemical & Engineering News
Vol. 68, p. 5, January 21, 1980

Joining-Up

This season joining-up set in, which is a crucial test of mapping. For when the natural lines run-on from one map-maker's ground into another's, then their maps, laid edge to edge, ought of course to correspond; and if not then one or both officers must be wrong. The readers will not be surprised to hear that hot disputes sometimes arise. But even between the most careful of workers, it is not wise to join-up along the edge of a map, for that has no visible existence in Nature. A better frontier is a stream, for that is visible on the ground, and positions can be fixed with ease.

Here are two legends, handed-down from the Heroic Age.

Scene: the old Cartography Office; speaker: an officer famous for his truculence—"Joinin'-up, eh? If I see trouble brewin', I just draw a h–ll of a fault along the whole confounded frontier, and cut-off every blessed line."

The other I had from A. H. Green. He had been telling, in Jermyn Street, how Dakyns and he, tracing the same line to their frontier, had actually met upon the ground, so precise had been their work. But one of the listeners slowly said: —"H'm, yes. I happen to know that country well, and I can tell you fellows that the place where Green and Dakyns met was just the only public house for five miles around."

E. Greenly
A Hand Through Time
Vol. 1, 129, 1927

Mississippi's Meanders

One of the Mississippi's oddest pecularities is that of shortening its length from time to time. If you will throw a long, pliant apple-paring over your shoulder, it will pretty fairly shape itself into an average section of the Mississippi River; that is, the nine- or ten-hundred miles stretching from Cairo, Illinois, southward to New Orleans, the same being wonderfully crooked, with a brief straight bit here and there at wide intervals. The two-hundred-mile stretch from Cairo northward to St. Louis is by no means so crooked, that being a rocky country that the river cannot cut much.

The water cuts the alluvial banks of the "lower" river into deep horseshoe curves; so deep, indeed, that in some places if you were to get ashore at one extremity of the horsehoe and walk across the next half or three-quarters of a mile, you could sit down and rest a couple of hours while your steamer was coming around the long elbow at a speed of ten miles an hour to take you on board again. When the river is rising fast, some scoundrel whose plantation is back in the country, and there-

fore of inferior value, has only to watch his chance, cut a little
gutter across the narrow neck of land some dark night, and turn
the water into it, and in a wonderfully short time a miracle has
happened: to wit, the whole Mississippi has taken possession
of that little ditch, and placed the countryman's plantation on its
bank (quadrupling its value), and that other party's formerly
valuable plantation finds itself away out yonder on a big island;
the old watercouse around it will soon shoal up, boats cannot
approach within ten miles of it, and down goes its value to a
fourth of its former worth. Watches are kept on those narrow
necks at needful times, and if a man happens to be caught
cutting a ditch across them, the chances are all against his ever
having another opportunity to cut a ditch.

Pray observe some of the effects of this ditching business.
Once there was a neck opposite Port Hudson, Louisiana,
which was only half a mile across in its narrowest place. You
could walk across there in fifteen minutes; but if you made the
journey around the cape on a raft, you traveled thirty-five miles
to accomplish the same thing. In 1722 the river darted through
that neck, deserted its old bed, and thus shortened itself thirty-
five miles. In the same way it shortened itself twenty-five
miles at Black Hawk Point in 1699. Below Red River Land-
ing, Raccourci cut-off was made (forty or fifty years ago, I
think). This shortened the river twenty-eight miles. In our
day, if you travel by river from the southernmost of these three
cut-offs to the northernmost, you go only seventy miles. To do
the same thing a hundred and seventy-six years ago, one had to
go a hundred and fifty-eight miles—a shortening of eighty-
eight miles in that trifling distance. At some forgotten time in
the past, cutoffs were made above Vidalia, Louisiana, at Island
92, at Island 84, and at Hale's Point. These shortened the
river, in the aggregate, seventy-seven miles.

Since my own day on the Mississippi, cut-offs have been
made at Hurricane Island, at Island 100, at Napoleon, Arkan-
sas, at Walnut Bend, and at Council Bend. These shortened
the river, in the aggregate, sixty-seven miles. In my own time
a cut-off was made at American Bend, which shortened the
river ten miles or more.

Therefore the Mississippi between Cairo and New Orleans

seventy-six years ago. It was eleven hundred and eighty after the cut-off of 1722. It was one thousand and forty after the American Bend cut-off. It has lost sixty-seven miles since. Consequently, its length is only nine hundred and seventy-three miles at present.

Now, if I wanted to be one of those ponderous scientiific people, and "let on" to prove what had occurred in the remote past by what had occurred in a given time in the recent past, or what will occur in the far future by what has occurred in late years, what an opportunity is here! Geology never had such a chance, nor such exact data to argue from! Nor "development of species," either! Glacial epochs are great things, but they are vague—vague. Please observe:

In the space of one hundred and seventy-six years, the Lower Mississippi has shortened itself two hundred and forty-two miles. That is an average of a trifle over one mile and a third per year. Therefore, any calm person, who is not blind or idiotic, can see that in the old Olitic Silurian Period, just a million years ago next November, the Lower Mississippi River was upward of one million three hundred thousand miles long, and stuck out over the Gulf of Mexico like a fishing-rod. And by the same token any person can see that seven hundred and forty-two years from now the Lower Mississippi will be only a mile and three-quarters long, and Cairo and New Orleans will have joined their streets together, and be plodding comfortably along under a single mayor and a mutual board of aldermen. There is something fascinating about science. One get such wholesale returns of conjecture out of such a trifling investment of fact.

Mark Twain
Life on the Mississippi
Harper & Row, NY, 1874

The Ancient Swift Winged Bird
of the Gobi Desert

"The beginning of the whole thing was a radiator orna-
ment (Fig. 1) that I designed and had cast for my car," wrote
Dr. C. B. Fraser.   When, in August 1926, his colleague at
Cornell University, Dr. L. W. Sharp, accompanied Dr. Fraser
on a ten-day automobile trip through the Adirondacks, he saw
the bird.   Amused at its appearance, he suggested writing a
short burlesque life history of the bird to send to a mutual
friend.   So in the evenings, at little country hotels in the Adirondacks, they wrote some 25 pages of what later became the
Monograph.   The thing grew.

Fig. 1.   *Eoörnis* resting on a rocky pinnacle, supported by the rapid
vibration of the wings.

In December, it was announced that a Professor A. C.
Fotheringham of New Zealand would address a meeting of the
faculty and graduate students of the various Biological Departments of Cornell University.   For an hour and twenty minutes,

Dr. Fraser, wearing a false beard, gave a lecture on *Eoörnis pterovelox gobiensis*—illustrated with slides projected by Dr. Sharp. "Quite a number of the audience failed to realize that the lecture was a fake until right near the end."

The joke took hold. The lecture was repeated, by request, at several conventions. With the financial support of friends who founded the Eoörnis Institute of America, the 34-page Monograph was published. As a burlesque of a scientific thesis, the Monograph is correct to the last detail of makeup, typography, and illustration (Fig. 2). The entire work is imaginary, in this sense: Though no one organism ever could display *all* the scientific features ascribed to *Eoörnis*, these features are actually true of some organism. The burlesque lies in the careful misstatement of scientific fact, and the appreciation of the burlesque presupposes an understanding of what the real facts should be. It is high tribute to the authors' skill and labor that such an extended spoof hangs together and sustains one's interest. The following excerpts, while from the less technical parts, may allow one to savor the Monograph.

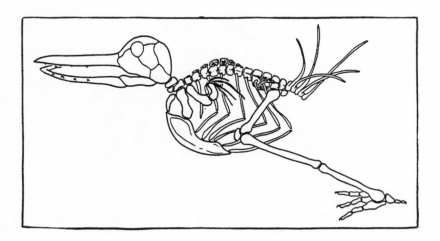

Fig. 2. The skeleton of *Eoörnis* (one side only is shown). Note particularly the dorsal ridge, the finger bones in the wing, and the unique tail support. The positions of the lumbar ganglion and the adrenal gland.

Eoörnis
pterovelox gobiensis

A Monograph by

Augustus C. Fotheringham
ScD (Cantab.), FRGS

Professor of Ornithological Palaeontology
in the University of New Zealand
Invercargill, New Zealand

This paper records what is probably the outstanding accomplishment of the British Museum's recent expedition in the Gobi Desert under the able leadership of Brigadier-General Sir Cecil Wemyss-Cholmondeley, Bart., of Buighleigh Gables, Hants, Herts, Straffordshire, England. The widespread controversy that has agitated scientific minds of recent years about the nature and habits, nay, even the very existence, of the remarkable organism that forms the subject of this contribution developed such importance that, at the insistance of Sir Cecil Wemyss-Cholmondeley, the British Museum authorized and supported the expedition. The author was honored with an invitation to accept the scientific directorship of the expedition, and was much pleased by the prospect of being able to settle once and for all the long-continued controversy, and, if possible, to find confirmation for the views he has so long championed in the face of unremitting and bitter opposition....

Nomenclature

The name by which the organism has long been known is that given it by Linnaeus: *Micropteron asiatica*. Since Linnaeus definitely states that he did not actually see a specimen, and since he further expressed doubt about the credibility of reports concerning it, the author has ventured to select as a more appropriate name that suggested in an obscure Persian journal by his esteemed colleagues Francois Villon and Sugamora, *Eoönoris pterovelox gobiensis*....

History

What is probably the earliest reference to *Eoörnis* appears in certain of the Cro-Magnon picture writings found in the caves of the Dordogne in France. These include undoubted representations of the organism in its characteristic resting attitude, invariably accompanied by a symbol hitherto undeciphered, but which is now known to represent music.

By an unusual bit of good fortune the expedition returned from the Gobi Desert just as another of England's recent expeditions, namely that to the tomb of Tutankamen, arrived in Cairo with its magnificient collections. Among the articles discovered in the tomb and placed in the author's hands for study were three amulets, each depicting the bird, again in the peculiar resting pose....

It would indeed be strange if among the writings of Marco Polo there should not be found some reference to *Eoörnis*, and in truth this famous traveler gives a very complete description of the bird itself, together with remarks as to its habits which we have been able to confirm *in toto*.

Psychology

The psychological behavior of *Eoörnis* is in many respects very remarkable. In the first place, the species is gregarious and gives evidence of some degree of social organization. Indications of this are seen in the Sumerian arrow form of flight, with the apparent selection of leaders that fly at the head of the arrow; the uniform monogamy; the coordinating effect of the vocal note; and the otherwise inexplicable grouping of nests, with corresponding inviolability of property rights.

The presence of both cerebral and lumbar ganglia gave rise to the surmise that the birds were enabled thereby to reason both *a priori* and *a posteriori*, as has been claimed for the dinosaur and allied species. Confirmation of the surmise was provided in crucial experiment. A full-grown adult that has been deprived of food for a week was placed between two piles of sand-fleas of exactly equal weight on a level rock in dull weather, so that illumination was equal on all sides. The bird was placed exactly midway between the piles and facing one of them, care being taken to see that the bird was aware of the presence of both piles. It was the expectation of the author in conducting this experiment that the bird would feed from one pile or the other, indicating which type of reasoning controlled its action. What was his surprise to find that the two ganglia were of equal functional potential: *a priori* and *a posteriori* reasoning in this artificially controlled environment exactly balanced, and after seven days of anguished mental struggle the bird died of starvation. It was unable to decide from which pile to eat first....

C. B. Fraser and **L. W. Sharp**
writing as **Augustus C. Fotheringham**
London, *The Buighleigh Press,* 1938

Gottlieb Laser

It is part of the folklore of modern science that the word *laser* is an acronym. My researches, which indicate that this is not so, were sparked off when browsing in a file of *The Times* for 1871. I noticed a despatch from their war correspondent, Arthur Jenn (originator of the modern expression "the gen" meaning the real truth) datelined "Before Paris." Further investigation revealed the circumstance of Gothick drama in which the laser was discovered.

Gottlieb Laser, born in 1830, was heir to enormous estates in East Prussia and to a brewing firm, Laser AG. He was a scientific dilettante with interests in geology and military engineering; he was also an amateur lapidary. When visiting German East Africa, he fell in love with a beautiful African woman, Iwo, whom he married and took back to Germany with him. Her earlobes were, by tribal custom, pierced by apertures 11 mm in diameter, and Laser, a very sentimental man, sought long for suitable ornaments for these features. At last he found, in Sierra Leone, a prodigious ruby and proceeded to work this into a right-circular cylinder 5-cm long. This, a true labor of love, took seven years, being protracted by constant experimentation to ensure a secure grip without discomfort.

When the Franco-Prussian war broke out, Laser became famous by inventing a war gas that completely neutralized the perfumes used by Frenchwomen and consequently profoundly depressed the morale of the French army. For this he was given an honorary colonelcy in the Prussian Guard.

During the siege of Paris he was invited to visit the German lines as a VIP. His loving wife, wishing to protect him by occult powers, gave him her precious ruby as a charm, lashing it to the frogging of his tunic. Laser also took with him another dear possession, a silver stein mug that was unusual in that its base was hemispherical rather than flat, a scientific attempt to avoid undue wastage of beer by surface tension in reentrant angles.

At dinner one night outside his tent, Laser was approached by Jenn, who asked if the Germans would be defeated by French air power. (A balloon from beleaguered Paris was passing overhead.) The good Prussian bellowed "Donner und Blitzen!" and banged his silver stein onto the table so forcefully that the ruby rod slipped from its lashings and fell into the beer. A press photographer wished to record the the scene, but Laser's face had gone a deep nonactinic red in his emotion and the photographer had only wet collodion plates and an $f/16$ rapid rectilinear lens. Accordingly, to ensure enough light, he brought his tray of flash powder close to Laser, almost over his stein, and fired the flash.

The opposing armies, engaged in the desultory firing, were stunned to see a brilliant shaft of light, pencil-thin and straight as a linear equation, reach up and strike the balloon, which burst with a roar into a crown of flame, dashing the intrepid aeronaut to his death the field below. There was no more firing that night.

In the early hours of the morning, a secret meeting between the French and German general staffs, convened in a nearby farmhouse under flags of truce, agreed the further development of the laser should be abandoned in the interest of humanity. Laser was much relieved, for he had no idea how it had worked.

Of course, the ruby had been pumped by the photographer's flashlight, focused by the concave base of the stein. The violently agitated surface of the beer no doubt introduced a measure of Q-switching, accounting for the immense power of the beam.

Laser expired soon after, grieved by the loss of his magnificent moustache, incinerated by the flash powder. The ruby was long preserved at his home in East Prussia, but disappeared when it was overrun by the Russians in Hitler's War. It is significant, however, that Schawlow and Townes shared their Nobel Prize for the discovery with a Russian.

We shall probably never know where the ruby went, but one cannot help wondering.

Simplicius (**James Dyson**)
NPL News
Vol. 209, 1967; Sept. 21, p. 9

Acknowledgments

The editor and publisher wish to thank the following publishers and individuals who have kindly given permission for the use of copyrighted material:

NBS Standard: US Dept. of Commerce, National Bureau of Standards, 1980 (p. 1); Sir Arthur Eddington: *The Nature of the Physical World*, Cambridge University Press, Cambridge, UK, 1928 (p. 2); Marilyn T. Kocher: *Physics Today*, May 1978 (p. 3); October 1979 (p. 163); R. W. Wood: *Contemptuary Science* (pp. 3–4); Jagdish Mehra: *The Physicist's Conception of Nature*, Reidel, Dordrecht, Holland, 1973 (pp. 4–6); R. H. March: *Physics for Poets*, McGraw-Hill, New York, 1970 (p. 6); Walter Elsasser: *Memoirs of a Physicist in the Atomic Age*, Science History Publications (Neale Watson Academic Publications), New York, 1978 (pp. 6–7); Allan Holden, Bell Laboratories, Murray Hill, NJ (pp. 7–8); A. P. French (pp. 8–9); Milton Eisenhower (pp. 9–10); Richard T. Cox, The Johns Hopkins University, 1981 (p. 10); M. Minnaert: *De Sterrekunde en de Mensheid*, Servire BV, Katwijk, 1946 (pp. 11–12); Patrick Moore: *Can You Speak Venusian? A Guide to the Independent Thinkers*, Norton, New York, 1972 (pp. 12–14); Joseph Hausen: *Journal of Chemical Education*, Vol. 35, 1958 (p. 15); Ralph E. Oesper: *The Human Side of Scientists*, University of Cincinnati Publications, 1975 (pp. 16–18); Peter Brain: *The Lancet*, Vol. 2, 1969 (pp. 18–22); Frank M. McMillan: *The Chain Straighteners*, Macmillan, London, 1981 (pp. 22–23); Sir Alexander Todd: *A Time To Remember: The Autobiography of a Chemist*, Cambridge University Press, Cambridge, UK, 1983 (pp. 23–27); M. D. Picard: *Grit and Clay*, Elsevier, Amsterdam, 1975 (pp. 28–30); Michael Ruse: *The Darwinian Revolution*, University of Chicago Press, Chicago, IL, 1979 (pp. 30–31); Peter Brown: *Computing Europe*, October, 31, 1974 (pp. 31–34); Nov. 6, 1974 (pp. 35–38); Ernest A. Meyer: *Microorganisms and Human Disease*, Appleton-Century-Crofts, New York, 1974 (p. 34); Kenneth C. Parkes, Richard C. Banks, and Robert W. Storer: *The Antic Alcid: An Anthology of the Auklet*, The American Ornithologists' Union, New York, 1983 (pp. 38–40 and p. 270–272); Kenneth J. Bell: *Heat Transfer Engineering*, Hemisphere Publishing Corporation, New York, Vol. 2 (3–4), 1981 (pp. 40–41).

Dan Nessett (pp. 59–60); John B. Haney: *AAUP Bulletin*, American Association of University Professors, February 1977 (pp. 61–68); Elizabeth M. Kerr: *AAUP Bulletin*, American Association of University Professors, Vol. 42, 1956 (pp. 68–69); Michael Frayn: *The Tin Men*, Collins, London, 1965 (pp. 69–75); A. J. Monty White: *Chemistry in Britain*, Vol. 13 (4), 1977 (pp. 75–79); Douglas E. Kidder, from *Verses Bright and Beautiful*, University of Bristol, 1983 (pp. 79–80 and 170–172); A. G. Foraker: *Journal of the American Medical Association*, Vol. 187, March 14, 1964 (pp. 81–83); Ian Rose: *Canadian Medical Association Journal*, Vol. 87, Dec. 8, 1962 (pp. 83–86) and *Medicarnage* (pp. 98–99); Carl Pacifico: *Chemical Engineering*, McGraw-Hill Inc., New York, November 20, 1967 (pp. 86–95); John Kearney (p. 95); Karl K. Darrow: *Bulletin of the American Physical Society*, Vol. 19 (1), 1944 (p. 96).

Richard H. Okenfuss (p. 97); Ruth Straus: *The Lancet*, December 26, 1959 (pp. 100–106); *The Percolator*, The Chemist's Club of New York, 1918 (pp. 107–108); Thomas R. Blackburn: condensed from *Science*, American Association for the Advancement of Science, Washington, DC, September 21, 1973 (pp. 108–113); H. A. Hoffman (pp. 113–118); Stanley A. Rudin: *Psychological Reports*, Vol. 5, 1959 (pp. 118–120).

K. M. Reese: *Chemical & Engineering News*, American Chemical Society, Washington, DC, Vol. 60, Aug. 23, 1982 (p. 121); Vol. 60, Jan. 4, 1982 (p. 122); October 29, 1979 (pp. 240–241); W. V. Sessions: *Journal of Chemical Education*, Vol. 3, 1926 (pp. 122–126); Robert Resnick: *Oersted Medalist Lecture*, AAPT-APS Meeting,

Anaheim, Dec., 1974 (pp. 126–127); Louis Berman (p. 128); *Journal of Chemical Education:* Vol. 2, 1925 and Vol. 3, 1926 (pp. 129–130); Vol. 24, 1947 (p. 170); W. D. Ian Rolfe: *Geological Howlers*, Geological Society of Glasgow (pp. 130–132); D. A. Davenport, M. Howe-Grant, and V. Srinivasan: *Journal of Chemical Education*, 56, 1979 (pp. 132–136); *Chemical Engineering:* McGraw-Hill Inc., New York, Jan. 20, 1964 (p. 136–138); J. E. Ware and R.G. Williams: *Journal of Medical Education*, Vol. 50, Feb., 1975 (pp. 139–141); Christopher T. Russell: *EOS, American Geophysical Union*, Washington, DC, 59 (3), 1978 (pp. 164–168); 60 (49), Dec., 1979 (pp. 141–145); William B. Jensen (pp. 145–146); G. Alex Mills: *Humor Among Chemists*, Univ. of Delaware, 1962 (pp. 146–150).

David J. Rose: *Technology Review*, Vol. 79, 2, 1976 (pp. 155–156); Mary Burwell (p. 157); *Observatory:* June 1931 (p. 157); Dec. 1935 (p. 160); Ms. Otto R. Frisch (pp. 161–162) *Proc. Chemical Society*, pp. 8–10, Jan., 1963 (p. 173); Dave Barry: *The Miami Herald*, 1982; Hubert F. Goenner: *Physics Today*, May, 1982 (pp. 177–178).

Harvey Lynn, Jr.: *Operations Research*, Operations Research Society of America, Vol. 4, 1956 (pp. 179–186); F. N. Young and Sears Crowell: *AIBS Bulletin*, American Institute of Biological Sciences, Washington, DC, Vol. 6, 1956 (pp. 186–190); Blanche Descartes: *Eureka* (No. 13), October, 1950 (pp. 197–198); Samuel B. Hand and Arthur S. Kunin: *Journal of the American Medical Association*, 214, December 28, 1970 (pp. 199–205); Richard W. Siegel (p. 206); E. G. Gray: *Nature*, Macmillan Journals Limited, Washington, DC, Vol. 304, August 25, 1983 (pp. 207–208); I. N. Dubin: *Perspectives in Biology and Medicine*, University of Chicago Press, Vol. 22, Autumn 1978 (pp. 208–215); Larry S. Liebovitch: *Quarterly Journal of the Royal Astronomical Society*, Vol. 15, 1974 (pp. 215–220); J. D. Pye: *Nature*, Macmillan Journals Limited, Washington, DC, Vol. 218, May 25, 1968 (pp. 220–221); Benjamin A. Horwitz: *Chemical Engineering*, McGraw-Hill, New York, August 9, 1982 (pp. 224–227); Julius S. Greenstein: *Canadian Medical Association Journal*, Vol. 93, Dec. 1965 (pp. 227–228); Edward Schreiber and Orson L. Anderson: *Science*, American Association for the Advancement of Science, Washington, DC, Vol. 168, 1970 (pp. 229–230); Jack E. Bissey: *Journal of Chemical Education*, Vol. 46, August, 1969 (pp. 232–234); Stephen Leacock: *Winnowed Wisdom*, Dodd, Mead, New York, 1926 (pp. 235–236); Charles W. Trigg: *Journal of Recreational Mathematics*, Vol. 13, 1980–1981 (p. 236); E. Scott Barr: *American Journal of Physics*, Vol. 23, Dec., 1955 (p. 237); *BioScience:* American Institute of Biological Sciences, Vol. 34, April, 1984 (pp. 238–240); Robert J. Nash: *Chemtech*, October, 1979 (pp. 242–244).

Frederic Brown: *Angels and Space Ships*, Dutton, New York, 1954 (pp. 245–246); Ralph A. Lewin: *The Biology of Algae and Other Verses*, Univ. Press of America, Washington, DC, 1981 (pp. 246–249); J. Earl Roberts: *Clinical Pediatrics*, J. B. Lippincott, Philadelphia, Vol. 3 (6), June, 1964 (pp. 255–266); A. L. Taylor (p. 269).

David Rostoker (pp. 269–270); *International Stop Continental Drift Society Newsletter:* Vol. 3, (3), January, 1981 (p. 270); *Nature:* Macmillan Journals Ltd, Washington, DC, Vol. 290, April, 1981 (p. 274); John Archibald Wyler: Wilmington Press, Harvard College Observatory (p. 277); Frank T. Gucker, Jr.: *Chemical Bulletin*, June, 1932 (p. 277); Norman Radin (pp. 279–284); D. K. Blackmore, D. G. Owen, and C. M. Young: *The Veterinary Record*, Vol. 90 (14), April 1, 1972 (pp. 285–291); American National Standards Institute, 1984 (pp. 292–300); Axel Leijonhufvud, condensed from *Western Economic Journal*, Vol. 11, Sept., 1973 (pp. 301–304).

Richard C. Sheridan: *Chemtech*, American Chemical Society, Washington, DC, April, 1976 (pp. 305–307); Adrian Berry: *High Skies and Yellow Rain, Daily Telegraph*, London, 1983 (pp. 307–310); Henry Petroski: *Chicago Sun-Times*, News Group Chicago Inc., 1986 (pp. 311–313); M. Ian Phillips: *Science*, American Association for the Advancement of Science, Washington, DC, January 17, 1975 (p. 314); R. V. Jones: *Bulletin of the Institute of Physics*, June, 1957 (p. 315); John Waugh (pp. 319–320); Monte Throdahl: *Chemical & Engineering News*, American Chemical Society, Washington, DC, Vol. 68, January 21, 1980 (pp. 320–321); James Dyson: *NPL News*, Optical Society of America, Vol. 209, 1967 (pp. 329–331).

Index*

*Names in italics are those of persons quoted as authors.